The Neuropsychology of Individual Differences

The Neuropsychology of Individual Differences

Philip A. Vernon
Department of Psychology
The University of Western Ontario
Social Science Centre
London, Ontario, Canada

Academic Press
San Diego   New York   Boston   London   Sydney   Tokyo   Toronto

Copyright © 1994 by ACADEMIC PRESS, INC.
All Rights Reserved.
No part of this publication may be reproduced or transmitted in any form or by any means, electronic or mechanical, including photocopy, recording, or any information storage and retrieval system, without permission in writing from the publisher.

Academic Press, Inc.
A Division of Harcourt Brace & Company
525 B Street, Suite 1900, San Diego, California 92101-4495

United Kingdom Edition published by
Academic Press Limited
24-28 Oval Road, London NWl 7DX

Library of Congress Cataloging-in-Publication Data

The Neuropsychology of individual differences / edited by Philip A.
Vernon.
     p.   cm.
   Includes bibliographical references and index.
   ISBN 0- 12-718670-0
   1. Neuropsychology.   2. Personality--Physiological aspects.
3. Individual differences--Physiological aspects.   I. Vernon,
Philip A.,
   [DNLM: 1. Neuropsychology.   2. Individuality.   3. Personality
Development.   WL 103 N4938 1994]
QP360.N4958   1994
612.8--dc2O
DNLM/DLC
for Library of Congress   94-7647
                                                                      CIP

Contents

Section I   Mental Abilities

3 Intelligence and the Brain
Douglas K. Detterman

4 The Neuropsychology of Sex-Related Differences in Brain and Specific Abilities: Hormones, Developmental Dynamics, and New Paradigm
Helmuth Nyborg

Section II   Information Processing and Memory

5 Neuropsychological Models of Information Processing: A Framework for Evaluation

W. Grant Willis and Andrew D. Aspel

Section III   Personality

9 Neuropsychology of Temperament
Jose J. Gonzalez, George W. Hynd, and Roy P. Martin

Contributors

Numbers in parentheses indicate the pages on which the authors' contributions begin.

Andrew D. Aspel (117), Chafee Social Science Center, Department of Psychology, University of Rhode Island, Kingston, Rhode Island 02881

Douglas K. Detterman (35), Department of Psychology, Case Western Reserve University, Cleveland, Ohio 44106

Hans Eysenck (151), Institute of Psychiatry, University of London, Denmark Hill, London SE5 8AF, United Kingdom

Gerald Goldstein (209), Department of Psychology, Highland Drive VA Medical Center, Pittsburgh, Pennsylvania 15206

Jose J. Gonzalez (235), Department of Educational Psychology, University of Georgia, Athens, Georgia 30602

Marion I. S. Huettner (9), Medical Associates Clinic, P.C., Dubuque, Iowa 52002

George W. Hynd (235), Department of Educational Psychology, University of Georgia, Athens, Georgia 30602

Daniel W. Kee (135), Department of Psychology, California State University at Fullerton, Fullerton, California 92634

Roy P. Martin (235), Department of Educational Psychology, University of Georgia, Athens, Georgia 30602

Helmuth Nyborg (59), PNE Research Center, Institute of Psychology, University of Aarhus, DK-8240 Risskov, Denmark

Mitchell Rabinowitz (135), Department of Educational Psychology, Fordham University, New York, New York 10023

Philip A. Vernon (3), Department of Psychology, The University of Western Ontario, London, Ontario, Canada N6A 5C2

W. Grant Willis (117), Department of Psychology, University of Rhode Island, Kingston, Rhode Island 02881

Preface

Individual differences in abilities, personality traits, and many other human behaviors have been of interest to psychologists for as long as psychology has been a field of scientific investigation. Moreover, some of the earliest researchers who examined individual differences sought to explain them in terms of underlying biological mechanisms. With the recent growth in neuroscience research, and increasing interest in the neuropsychological foundations of personality and behavior, a greater awareness of the foundations of individual differences has been made possible. This volume presents an overview of this research.

Although no single volume can discuss every aspect of individual differences, chapters have been chosen to reflect some of the most common areas of inquiry and research. The topics discussed here will illustrate the important contribution that the neuropsychological approach can make to our understanding of the nature and sources of individual differences. The volume is divided into three sections, encompassing individual differences in abilities, information processing and memory, and normal and abnormal personality.

Section I begins with an introductory chapter describing two powerful and useful behavioral genetic procedures, genetic correlation and group heritability. These procedures serve as examples of what behavioral genetic methodology has to offer and should be of considerable interest to researchers in neuropsychology. In Chapter 2, Huettner discusses language and reading development. Chapter 3 contains a discussion of human intelligence by Detterman. Completing the section on abilities, Chapter 4 is a provocative examination of the neuropsychology of sex-related differences in specific abilities by Nyborg.

Section II contains chapters by Willis and Aspel on neuropsychological models of information-processing (Chapter 5) and a discussion of individual differences in memory by Rabinowitz and Kee (Chapter 6).

In the final section, Hans Eysenck addresses biological foundations of personality in Chapter 7. Chapter 8, by Goldstein, discusses psychopathology and schizophrenia. The final chapter, by Gonzales, Hynd, and Martin, focuses on the neuropsychology of temperament.

Thanks are due to the authors, both for the quality of their contributions and for their exemplary adherence to deadlines! Thanks also to Nikki Fine and Diane Scott at Academic Press for their assistance and patience during the completion of this project. Finally, thanks to several colleagues and neuropsychologists who commented on different chapters during the preparation of this volume.

Philip A. Vernon

Section I
Mental Abilities

1

Introduction to the Neuropsychology of Individual Differences

Contributions of Behavioral Genetics

Philip A. Vernon

I. INTRODUCTION

Individual differences in abilities, personality traits, and many other human behaviors have been of interest to psychologists for as long as psychology has been a field of scientific investigation. Moreover, some of the earliest researchers who examined individual differences sought to explain them in terms of underlying biological mechanisms.

In this volume, individual differences in abilities, information processing, memory, and normal and abnormal personality are examined from a neuropsychological perspective. In Chapter 2, Huettner discusses language and reading development. Chapter 3 contains a discussion of human intelligence by Detterman. Completing the section on abilities, Chapter 4 by Nyborg is a provocative examination of the neuropsychology of sex-related differences in specific abilities.

Section II comprises a chapter by Willis and Aspel on neuropsychological models of information processing (Chapter 5) and a discussion of individual differences in memory by Rabinowitz and Kee (Chapter 6).

In Section III, Hans Eysenck addresses biological foundations of personality in Chapter 7. In Chapter 8 Goldstein discusses psychopathology and schizophrenia. The final chapter, by Gonzalez, Hynd, and Martin, focuses on the neuropsychology of temperament.

This volume does not claim to address all aspects of individual differences that have been examined from a neuropsychological perspective. The topics discussed here will attempt to illustrate the important contribution that the neuropsychological approach can make to our understanding of the nature and sources of individual differences in a wide range of behaviors.

One topic that is not addressed in the other chapters concerns recent developments in the field of behavioral genetics, and the advantages of behavioral genetic methodology, which sometimes goes above and beyond traditional approaches to the study of individual differences. Too often, behavioral genetics is perceived as doing little more than estimating heritability coefficients (not that this may not be a useful enterprise in its own right), but recent advances in the field have made it possible to use twin and other genetically informative data to address questions about the genetic and environmental bases of individual differences that simply cannot be answered by any other means. Two examples—genetic correlations and group heritability—will be used to illustrate some of these advances. Additional examples can be found in an excellent volume by Neale and Cardon (1992).

A. Genetic Correlation

A phenotypic correlation between two variables is the observed correlation, typically a Pearson r in the case of two quantitative variables, with which we are all familiar. Behavioral genetic methodology allows one to estimate the extent to which such an observed phenotypic correlation represents (1) correlated *genetic* factors, (2) correlated *environmental* factors, (3) or both. This is done by computing genetic correlations.

Consider the correlation between intelligence quotient (IQ) scores, as measured by a standard test of intelligence, and speed of information processing, as measured by one or another reaction time (RT) test. Phenotypic correlations between IQs and RTs typically range between about $-.20$ and $-.40$ (Vernon, 1987), indicating that higher IQ scores are associated with faster speed of information processing. Given an IQ–RT correlation of, say, $-.30$, the question may be asked whether IQ scores and speed of processing are correlated because both are dependent upon the same *genes*—that is, do the same biological or physiological factors operate on both intelligence and on speed of information processing—or are there certain *environmental* factors that influence both a person's IQ and the speed with which the person can process different kinds of information, or do both work together?

Multivariate analyses of twin or other genetically informative data, typically employing LISREL (Joreskog & Sorbom, 1986) or Mx (Neale, 1991), can answer these questions by examining the twin cross correlations for the variables under investigation. Cross-correlations, in the present example, are obtained by correlating the IQ score of one member of a twin-pair with the RT of the other twin. This is done separately for monozygotic (MZ) and dizygotic (DZ) twin-pairs. To the extent that the MZ cross-correlation is greater than the DZ cross-correlation, the phenotypic correlation between the two variables is dependent upon correlated genetic factors. If the MZ and DZ cross-correlations are equal, then common environmental factors are entirely responsible for the phenotypic correlation. In a similar fashion to estimating the heritability of a trait by doubling the difference between MZ and DZ intraclass correlations, the genetic correlation between two variables can be estimated by doubling the difference between MZ and DZ cross-correlations and dividing the difference by the square root of the product of the variables' heritabilities.

A study by Baker, Vernon, and Ho (1991) reported phenotypic and genetic correlations between verbal and performance IQ scores and general speed-of-processing factor scores derived from eight RT measures administered to samples of adult MZ and DZ twins. Phenotypic correlations between RTs and verbal and performance IQs were both −.59, and LISREL analyses showed that a very sizable proportion of these correlations was attributable to correlated genetic factors: the genetic correlation between verbal IQ and RT was, in fact, 1.00; that between performance IQs and RTs was .92. These results indicate that genetic variations that lead to faster information processing are highly associated with genetic variations important to high IQ scores. From a neuropsychological perspective, the results also suggest that the search for biological factors that correlate both with intelligence and with speed of information processing should be a fruitful endeavor. Indeed, some results have already borne this out (see Chapter 3 by Detterman, this volume).

B. Group Heritability

The concept of group *heritability* is one that should be of particular relevance to neuropsychologists interested in the etiology of the difference between individuals who fall at one or another extreme of the distribution of some trait and the rest of the population. Consider mental retardation as an example. Is it the case that mildly or severely retarded individuals fall at the low end of the distribution of normal intelligence, and that the same environmental and genetic factors that operate on the rest of the distribution also operate on their IQs? Or, do different environmental and genetic factors contribute to mental retardation in comparison to those that operate on the rest of the distribution, or is it a question of both factors? Estimates of group familiality (obtained from sibling data) and group heritability (obtained from MZ twin data) can address these questions.

Group familiality for mental retardation (or any other quantitative trait or disorder) is estimated by comparing the mean IQ score of diagnosed retarded individuals to the mean of their siblings (DeFries & Fulker, 1985, 1988; Roberts, 1952). The degree to which the mean of the siblings regresses up towards the population mean indicates the degree to which familial factors are *unimportant* in the etiology of mental retardation. In contrast, if the mean of the siblings is close to that of the probands, familial factors are implicated. Roberts (1952) showed that siblings of mildly retarded individuals were themselves also below average in IQ, indicating that mild retardation is familial. On the other hand, siblings of severely retarded individuals were not below average in IQ, indicating that severe mental retardation is not familial and that its etiology is different from the rest of the IQ distribution.

A further comparison can be made between the estimates of *group* familiality and traditional estimates of *individual* familiality (i.e., estimates of the extent to which individual differences in the general population are attributable to familial factors). If group familiality differs from individual familiality—as it does in the case of severe mental retardation—this again indicates a difference in etiology. If group and individual familiality are similar or equal in magnitude—as they are in the case of mild mental retardation—this *may* indicate that mild retardation represents the low end of the normal distribution of intelligence, although it is also still possible that those factors that contribute to mild mental retardation are different from the factors that contribute to individual differences in intelligence in the normal range.

If twins rather than nontwin siblings are studied, estimates of group *heritability* can be obtained. The group heritability of a trait indicates the extent to which the difference between individuals who score at the extreme on the trait and the population mean can be attributed to genetic differences. It is found by comparing the degree to which the mean scores of MZ and DZ cotwins of the extreme scorers regress back to the population mean. If MZ and DZ cotwins regress to the population mean by the same amount, the group heritability of the trait is zero, and none of the difference between the extreme scorers and the population mean can be attributed to genetic factors. On the other hand, if MZ cotwins obtain the same mean as their extreme-scoring twins, whereas DZ cotwins regress halfway to the population mean in comparison to their extreme-scoring twins, the group heritability of the trait would be 1.0.

Plomin (1991) presents an analysis of twin data from a study of reading disabilities by DeFries and Fulker (1985). In this study, the mean discriminant function scores obtained from a battery of reading, perceptual speed, and memory tests of MZ and DZ twin-pairs, at least one of whom had a diagnosed reading disability, were compared to the mean of nontwin controls. Expressed as standardized deviations from the mean of the controls, set at 0, the diagnosed reading-disabled twins (MZs and DZs combined) obtained a mean of -2.8; the means of the MZ and DZ twins of these diagnosed probands were -2.0 and -1.6, respectively.

Estimates of group familiality from these data are found from the ratios of 2.0/2.8 = .71 for the MZs and of 1.6/2.8 = .57 for the DZs. Group heritability is estimated by doubling the difference between these two estimates of group familiality [i.e., 2(.71 − .57) = .28]. This latter estimate is interpreted as showing that 28% of the difference between the diagnosed reading-disabled probands and the general population is attributable to genetic factors. Common environmental factors account for .71 − .28, or 43% of the difference. Unique or nonshared environmental factors account for the remaining 29% of the difference. Plomin (1991) also reports that the heritability of individual differences in reading ability is approximately .50. The difference between this individual heritability of .50 and the group heritability for reading disabilities of .28 indicates that the genetic etiology of reading disabilities is different from that of the normal distribution of reading scores.

These examples of group familiality and group heritability should illustrate the potential of the methodology when applied to a comparison of normal and abnormal behaviors or normal and abnormal performance on any quantitative trait. Estimating the extent to which the genetic and environmental etiologies of extreme scores on a trait are the same or different from those of individual differences in the normal distribution of the trait may serve as a useful starting point in many neuropsychological as well as other investigations.

Genetic correlations and group familiality or group heritability are but two examples of the application of behavioral genetic methodology to questions that would be of interest to neuropsychologists and others studying individual differences. Twin data (or data from adoptees and their adoptive and biological relatives) have many other useful applications. For example, in certain situations it is possible to determine the direction of causation between variables (e.g., temperament and alcoholism, or adverse life events and depression) by comparing MZ and DZ cross-covariances (Neale & Cardon, 1992). Great advances in the area of molecular genetics have also been made that will be applicable to the study of behavior and behavior disorders (Plomin & Rende, 1991). Behavioral genetics is a field that has gained considerable recognition in recent years and it is hoped that neuropsychologists and others will continue to avail themselves of the advantages that this field has to offer.

REFERENCES

Baker, L. A., Vernon, P. A., & Ho, H.-Z. (1991). The genetic correlation between intelligence and speed of information-processing. *Behavior Genetics, 21,* 351–367.

DeFries, J. C., & Fulker, D. W. (1985). Multiple regression analysis of twin data. *Behavior Genetics, 15,* 467–473.

DeFries, J. C., & Fulker, D. W. (1988). Multiple regression analysis of twin data: etiology of deviant scores versus individual differences. *Acta Geneticae Medicae et Gemellologiae, 37,* 205–216.

Joreskog, K. G., & Sorbom, D. (1986). *LISREL: Analysis of Linear Structural Relationships by the Method of Maximum Likelihood.* Chicago: National Educational Resources.

Neale, M. C. (1991). *Mx: Statistical Modeling.* Department of Human Genetics, Box 3 MCV, Richmond, VA 23298.

Neale, M. C., & Cardon, L. R. (1992). *Methodology for genetic studies of twins and families.* Kluwer, Dordrecht, The Netherlands. [NATO Advanced Science Institute Series D: Behavioural and Social Sciences, Vol. 67]

Plomin, R. (1991). Genetic risk factors and psychosocial disorders: Links between normal and abnormal. In M. Rutter & P. Casaer (Eds.), *Biological risk factors for psychosocial disorders.* Cambridge, England: Cambridge University Press.

Plomin, R., & Rende, R. (1991). Human behavioral genetics. *Annual Review of Psychology, 42,* 161–190.

Roberts, J. A. F. (1952). The genetics of mental deficiency. *Eugenics Review, 44,* 71–83.

Vernon, P. A. (Ed.). (1987). *Speed of information-processing and intelligence.* Norwood, NJ: Ablex.

2

Neuropsychology of Language and Reading Development

Marion I. S. Huettner

The quest for the neurological bases of language and reading is over 100 yr old (Kirk, 1983). Most of what is known about the neural bases of language and reading is based on clinical and research findings with adult samples. Based on the adult aphasia, alexia, and acquired dyslexia literature and Luria's conceptualization of language and reading as multidimensional, interactive functional systems, several contemporary neuroanatomical and neurolinguistic models of mature reading have been advanced (Huettner, 1989). However, in addition to failure to account for the role of attentional factors and subcortical and frontal structures in language and reading, contemporary models are increasingly criticized for their limited applicability to how developing brains acquire language and reading skills (Holmes-Bernstein & Waber, 1990; Huettner, 1989).

Aware of these limitations, the central purpose of this chapter is to review recent literature and issues relevant to the search for the neural bases of language and reading *development*. This quest necessarily includes examining current knowledge of both normal and dysfunctional language and reading in pediatric populations. Thus, after an overview of contemporary adult models of reading as the necessary background for this endeavor, normal language and reading development, based on neurolinguistic literature, along with prominent theories and model-building efforts will be discussed. Subsequently, a brief discussion of developmental speech–language and reading disorders, and the relationship between them will be presented. Relevant electrophysiological data and their contribution to understanding the neural basis of language and reading development will be highlighted. Finally, the contribution of studies of early focal brain lesions to building models of the functional organization of language and reading in children is discussed.

I. OVERVIEW OF ADULT MODELS OF LANGUAGE AND READING

Geschwind (1962, 1972) synthesized Wernicke's conceptualization of the central aphasias and Dejerine's intensive case studies of alexia with and without agraphia into a model that presupposes that written material is initially registered in the bilateral occipital lobes where individual strokes of print are integrated into known letters and words. This visual information is then conveyed to the angular gyrus where it is combined with input from other sensory modalities to permit cross-modal activities, such as grapheme–phoneme correspondence. From the angular gyrus, the information traverses to Wernicke's area, where linguistic–semantic comprehension occurs, and is then shared with Broca's area via the arcuate fasciculus to result in oral reading. This model suggests that when a word is heard, the primary auditory receiving area in the temporal lobe conveys the information to Wernicke's area where it is comprehended. If the word is to be spoken, the pattern is transmitted from Wernicke's to Broca's area where the articulatory form of the word is activated and passes to the motor area that controls movement of the speech musculature (Geschwind, 1972). Geschwind further writes that if a word is to be spelled, the auditory pattern travels from the primary auditory zone to Wernicke's area to the left angular gyrus, where it elicits the visual pattern of the word, which explains why patients with alexia with agraphia may also have spelling deficits and cannot recognize words spelled out loud (Mayeux & Kandel, 1985).

Bastian (1898) proposed a model of reading similar to the Geschwind–Dejerine model to account for alexia without agraphia with the exception of more active right hemisphere involvement. Although he retained the classic left hemisphere perisylvian language structures in his model, Bastian gave a more prominent role to what he called the "right visual word center," which was the right hemisphere homolog for the angular gyrus. According to Bastian's model, printed words are received by the right medial occipital lobe. The linguistic–semantic aspects of the words are conveyed via the corpus callosum to the left medial occipital lobe, while the nonlinguistic, visual-spatial elements of print are conveyed to the right visual word center and then, via the corpus callosum, to the angular gyrus in the left hemisphere. From the angular gyrus, the information goes to Wernicke's area and then, via the arcuate fasciculus, to Broca's area, as in the Geschwind–Dejerine model.

Neurolinguistic models of reading have been derived primarily from studies of neurolinguistic error patterns in acquired dyslexia (Newcombe, Phil, & Marshall, 1981). Neurolinguistic models of reading have typically been based on dual-route theories of reading that propose an indirect, phonological process based on knowledge of grapheme–phoneme correspondence rules and a more direct, orthographic process, which accesses word-specific knowledge (i.e., sight vocabulary). Based on error types that reflect which decoding strategy predominates, three principal subtypes have been identified. Surface dyslexics rely primarily on indirect, phonological processes for accessing word meaning, whereas phonological and deep dyslexics make more use of nonphonological visual processes (Hynd & Hynd, 1984).

The neurolinguistic model presented by Hynd and Hynd (1984) and discussed by Huettner (1989) proposed that the hypothesized neural network underlying impaired semantic access to what is read in surface dyslexia, as reflected by their poor reading comprehension, involves bypassing both the bilateral visual association zones in the occipital lobes and Wernicke's area. In other words, printed information registers in the primary visual cortex and travels directly to the angular gyrus and from the angular gyrus directly to Broca's area. In contrast, the impaired process in deep dyslexia is grapheme–phoneme conversion, suggesting that the angular gyrus is bypassed. The proposed neural pathway, beginning at the primary visual cortex, appears to go through the right occipital lobe where the ability to image words is important and travels directly to Wernicke's area, where comprehension occurs. They retain the ability to go from a familiar visual configuration and especially concrete, imaginal nouns, directly to meaning but have much more difficulty reading nonimaginal nonwords, verbs, and abstract nouns. They demonstrate fairly good comprehension of meaningful material secondary to reliance on contextual cues. The putative neural pathway in phonological dyslexia is quite similar to that proposed for deep dyslexia with the exception that the left visual association cortex is involved as phonological dyslexics are able to process letter strings.

II. NORMAL LANGUAGE AND READING DEVELOPMENT

A. Normal Speech–Language Development

Understanding brain mechanisms in normal speech–language and reading development is essential to understanding developmental speech–language and reading disorders and vice versa (Pennington, 1991). In the last 10–15 yr, developmental psycholinguistics has provided much information about the course of normal language acquisition. However, normative indices have not been documented for many aspects of language development, and there is wide individual variation (Allen, 1989). What follows is a discussion of what is known about normal language acquisition processes at four commonly accepted levels of linguistic functioning—phonology, syntax and morphology, semantics, and pragmatics. Given the limited normative data available for specific language and reading processes from infancy through adolescence, any comments about the neural bases of language and reading acquisition and functional systems at various ages remain speculative and in need of empirical validation.

1. Phonological Development Phonology refers to the sound structure of language, including phonemes and syllables and the rules governing how sounds are combined in a given language (Sawyer & Butler, 1991). Phonological development as used in this chapter will refer to the development of phoneme perception and production. According to classical adult models of language functioning, phonological processing occurs mainly in Wernicke's area in the left temporal lobe and phonological production in and around Broca's area in the frontal lobes.

Pennington (1991) notes that lateralization of phonological processes to the left hemisphere is evident quite early in development. Discrimination of speech sounds develops as early as 2–3 months of age (Eimas, Siqueland, Jusczak, & Vigorito, 1971). Furthermore, work by Patricia Kuhl, summarized by Adler (1992a), suggests that infants discriminate between sounds in their own language and sounds not in their native language by 6 months of age, based on imitative models provided by caretakers.

With respect to production, crying is considered by many to be the first stage of speech–language development. De Hirsch (1984) notes that Lieberman showed that the wave forms of an infant's cry have the same characteristics as adult sentences and goes on to describe how neonates use the physiological mechanisms of respiration and phonation for communicative purposes. For example, discomfort cries apparent at birth approximate to narrow-front vowels and are usually nasalized, while several weeks later, infants produce relaxed, open-back vowels during states of well-being. Babbling occurs between 4 and 10 months (de Hirsch, 1984; Pennington, 1991). De Hirsch (1984) writes that during the babbling phase, sound-producing movements are being repeated over and over and strong associations are forged between tactile kinesthetic impressions and auditory sensations. Primitive sound–movement schemata are established, which serve as the foundation for storing and retrieving auditory–motor patterns that underlie articulatory competence. In the echolalic stage that begins at about 10 months and continues through about 29 months, infants actively imitate sounds and, later, single- and multiword utterances.

Although research on acquisition of vowel production is limited, vowel acquisition typically occurs before consonant mastery, with 2-yr-old subjects producing vowels with 75% accuracy and 3-yr-old subjects with 93% accuracy (Stoel-Gammon, 1991). De Hirsch (1984) notes that well-mothered babies master half of the consonant sounds by the end of their first year of life. Stoel-Gammon (1991) indicates that the range of correct consonant production in normally developing 2-yr-old subjects is 41–90%, and that this range of individual differences declines substantially between 2 and 3 yr of age. Further strides in articulatory competence become apparent by 28 months of age with the ability to use auditory feedback to monitor the accuracy of one's own speech production (de Hirsch, 1984). Articulatory competence for all phonemes is not complete until about age 8 (Pennington, 1991), and there is steady growth in auditory monitoring of one's own speech through about 9 yr of age (de Hirsch, 1984). Pennington (1991) further notes that there are steady increases in articulatory speed and automaticity at least into adolescence.

Stoel-Gammon (1991) points out that despite wide-ranging individual differences in the rate of phonological development in toddlers, several common denominators characterize normal phonological development. These commonalities include that consonant development proceeds in a specified order (i.e., stops, nasals, and glides before liquids, affricates, and fricatives); the number of initial consonants tends to be larger than the number of final consonants; the number of different sounds produced by a child is positively correlated with his vocabulary size; and age at onset of meaningful speech is negatively correlated with lexicon size and phonetic inventory at 24 months. See Stoel-Gammon (1991) for a discussion of the relationship between these characteristics of normal phonological development in toddlers and aspects of phonological impairment. Allen (1989) identifies phonological impairment in preschoolers by the presence of sound substitutions, omissions, or distortions that are unusual, inconsistent, or unpredictable. She also notes that many phonologically impaired preschoolers can spontaneously produce or accurately imitate isolated consonant–vowel syllables that break down in longer sequences.

2. Syntactical and Morphological Development   Syntax refers to the rules governing the sequencing of words in phrases and sentences and the contribution of word order to meaning (Ryalls, Beland, & Joanette, 1990; Sawyer & Butler, 1991). Often considered a level of linguistic functioning between phonology and syntax, morphology refers to the smallest units of meaning in the internal structure of words (Ryalls, et al., 1991) and will be considered a subset of syntax in this chapter. Components of syntactical development as discussed in this chapter include mastery of plural, possessives, tense markers, pronoun usage, negation, suffixes and prefixes, active and passive forms, statement and question forms, and various word classes. The adult aphasia literature suggests that individuals with Broca's aphasia have syntactic production problems and, to a lesser degree, receptive appreciation for structural information encoded in word order, suggesting that frontal systems subserve both productive and receptive syntax.

The onset of syntactical development occurs between 12 and 24 months with the production of the first nonimitative two- and three-word utterances (de Hirsch, 1984). Between 24 and 30 months, when expressive vocabulary has exceeded 100 to 200 words, there is a marked increase in acquisition and production of morphological inflections (e.g., possessive "s"; past tense "-ed") and grammatical function words (e.g., pronouns, prepositions, articles, etc.) (Thal, 1991). Although onset of phrase acquisition can range from 10–44 months, by 3–3.5 yr of age, children typically speak in three- to four-word sentences (de Hirsch, 1984; Stoel-Gammon, 1991). Usually by age 5 yr, children have mastered all the grammatical structures they will ever use (de Hirsch, 1984); however, they will continue to grow in automaticity and fluidity throughout the elementary school years (Sawyer & Butler, 1991).

Impaired development in syntax is typically marked by omission of obligatory forms or misapplication of learned grammatical rules (Allen, 1989). Tallal, Curtiss, and Kaplan (1988) note that syntactic deficits may result from basic phonological impairments.

3. Semantic and Lexical Development   The semantic level of linguistic development refers to the meaning or content attached to words or phrases (Sawyer & Butler, 1991). Semantic development involves the onset of object naming and increases in vocabulary comprehension and production (Allen, 1989). The adult literature suggests a strong role for the angular gyrus in the left parietal–occipital cortex in object naming and, to a lesser degree, the right posterior hemisphere, especially for imaginal words (Huettner, 1989).

Between 12 and 16 months, infants begin to name objects, although single-word naming can range from 6–30 months (Thal, 1991). About 2–3 months earlier, they begin to comprehend names spoken by others. Smith and Sachs (1990) found rapid growth in comprehension, but not production, of verbs at 14–16 months of age in their sample of 12–19-month-old subjects. Between 16 and 24 months, when expressive vocabulary reaches about 50 words, lexical development accelerates dramatically during the classic "vocabulary spurt." Typically, comprehension and production of nouns precede that of relational words like verbs and adjectives, although there is some evidence to suggest that there may be more than one way lexical development proceeds. For instance, Goldfield and Reznick (1990) found that 13 of their 18 infant subjects displayed the classic vocabulary spurt during which 75% of the words learned were nouns, whereas the lexical development of the other five subjects occurred more gradually and involved a balance of nouns and other word classes. As noted in the overview of syntactic development, when expressive vocabulary reaches 100–200 words between 24 and 30 months, syntactical and lexical development overlap and may be timed by common mechanisms as there is tremendous growth in the use of various morphemes and function words. By 2 yr, expressive vocabulary consists of about 300 words with content words predominating and approaches 1000 words at 3 yr, with even larger receptive vocabularies at each age (Stoel-Gammon, 1991). Vocabulary development continues to grow in most persons throughout the years in formal education, and, in the case of educated persons, throughout one's lifetime.

Semantic impairments manifest as impoverished vocabulary, overuse of nonspecific forms, and difficulties comprehending abstract language (Allen, 1991). Tallal et al. (1988) comment that restricted semantic functioning may be delimited by impaired syntactic–morphological development, since semantics is encoded, in part, through syntax and morphology. Similarly, semantic functioning is related to phonological development as well (Sawyer & Butler, 1991).

4. Pragmatic Development   Pragmatics refers to paralinguistic, nonverbal aspects of language such as nonpropositional speech (i.e., conventional expressions, speech formulas, idioms, frozen metaphors, and expletives that have a cohesive, unitary, unanalyzed structure), comprehension and communication of intentionality, emotionality, prosody, and gesture, as well as knowledge of complex social conventions like turn-taking, initiating conversation, and topic maintenance (Allen, 1989; Van Lancker, 1987). There is some empirical and clinical evidence that these nonlinguistic functions of language are relegated to the right hemisphere (Huettner, 1989; Van Lancker, 1987) and involve subcortical, frontal, and possibly left hemisphere structures as well (Pennington, 1991; Van Lancker, 1987).

There is a large empirical literature documenting associations between various aspects of normal lexical development and aspects of gestural development and usage (Bates, Thal, Whitesell, Fenson, & Oakes, 1989). Thal (1991) highlights some of these associations. The age of 10 months marks the onset of word comprehension and of intentional communication using vocal and gestural modalities, as well as tool use. Similarly, de Hirsch (1984) writes that interpretation of intonation precedes comprehension of phonemic patterns. Between 12 and 20 months, first words and recognitory gestures (i.e., conventionalized gestures associated with objects like drinking from a cup or sniffing a flower that are used to recognize or classify objects) emerge. Between 20 and 24 months, multiword utterances emerge along with production of multistep gestural combinations (e.g., scooping pretend food from a plate with a utensil and pretending to eat it in a smooth, single action plan). At 28–30 months, at the same time they are beginning to learn rules for English syntax and morphology, children spontaneously reproduce conventional gestures from familiar scripts and routines (e.g., playing house) in their play. Along these same lines, Smith and Sachs (1990) studied the cognitive-conceptual basis for the emergence of verbs in early lexical development in 12–19-month-old infants and found that engaging in symbolic play sequences and the ability to consider others in the role of actor during play with objects was linked to increases in verb comprehension at 14–16 months. Deficits in the pragmatic aspects of language include violation of turn-taking, difficulty initiating conversation, and poor topic maintenance.

Despite the growing literature relating aspects of linguistic functioning to aspects of nonlinguistic functioning at various points in development, these relationships are not well understood (Tallal et al., 1988). Specifically, whether disruption of linguistic development is accompanied by concomitant impairments in nonlinguistic functioning has not been settled by the research.

B. Normal Reading Development

Fluent reading is a multicomponential endeavor that involves processing linguistic information at the phonological, syntactic, lexical, and discourse levels (Snyder & Downey, 1991), and it also recruits a host of visual and conceptual-comprehension processes as well. Children between 6 and 8 yr in the beginning stages of reading focus their efforts and attention on learning sound–letter correspondences, decoding print into sound, and blending the sounds into recognizable words (Snyder & Downey, 1991). Between 8 and 9 yrs of age, most normal readers master the decoding process well enough that they can apply it automatically to print and allocate more attention and effort to integrating it with meaning they get from the syntactical structure at the sentence and text levels, thus moving into fluent reading (Chall, 1987; Snyder & Downey, 1991). After fourth grade, text typically goes beyond what the reader knows from prior experience, requiring more world knowledge and sophisticated language and cognitive abilities (Chall, 1987).

Based on this general overview of normal reading development, relationships to various aspects of linguistic development can be seen. The relationships between normal reading development and phonological, syntactic, and semantic development will be further discussed.

1. Normal Reading Development and Phonological Development Although skilled reading requires visual-perceptual processes for letter recognition as well as for comprehension processes, many argue that word recognition (i.e., decoding) is the central recurrent component in reading (Pennington, 1991). Decoding in normal readers typically occurs by means of both phonological and orthographic coding mechanisms (Szeszulski & Manis, 1990). Phonological coding refers to knowledge of regular relationships between spelling patterns and sounds and is used by most fluent readers to analyze unfamiliar words (Berninger, 1990; Recht, Newby, & Caldwell, 1992). Orthographic coding is based on word-specific versus rule-governed codes and provides direct access to meaning for familiar words (Berninger, 1990; Szeszulski & Manis, 1990). Pennington (1991) summarizes evidence that visual word recognition and visual word-form processing depend closely on the alphabetic code in some way, linking orthographic processing inextricably to phonological processing.

Phonological awareness, which refers to the ability to auditorally segment words into sounds and syllables, blend sounds into words, rhyme, and play phonetic games, has been considered a language skill that has particular relevance for word reading (Blachman, 1991; Sawyer & Butler, 1991; Snyder & Downey, 1991). Auditory segmenting, the process by which words, syllables, and later phonemes are isolated from the speech stream, is critical to learning letter–sound correspondences, and is acquired between 3 and 7 yr (Sawyer & Butler, 1991). Four and 5 yr olds can adequately segment words into onset (beginning consonant or consonant cluster) and rime (vowel and any succeeding consonant) (Snowling, 1991). Snowling (1991) further discusses evidence suggesting that short-term memory difficulties interfere with phonic blending during decoding and possibly with holding partially decoded words in memory while they are compared with pronunciations of words retrieved from long-term memory. Snyder and Downey (1991) note that how phonological skills relate to reading comprehension at different ages in the elementary years is not well explicated in the research.

The ability to use orthographic information automatically occurs early in the reading process, with little further development between 8.5 yr through college (Zecker, 1991).

Szeszulski and Manis (1990) reported individual differences in reliance on phonological and orthographic coding for word recognition in their normal controls. Only 12.5% of the normal controls were below 2.25 standard deviations from the mean on measures of phonological and orthographic coding, while 70% of those with at least one dyslexic parent fell below 2.25 standard deviations from the mean.

2. Normal Reading Development and Semantic and Syntactic Development

Word-retrieval skill may be one link between semantic development and normal reading acquisition. While naming ability is significantly related to early reading development, little is known about how naming ability relates to reading at later age levels (Snyder & Downey, 1991). There is some evidence that use of context clues to activate meaning and to assist word recognition decreases with acquisition of better reading skills (Snyder & Downey, 1991).

Narrative discourse is a higher order reading skill intimately related to reading comprehension. Narrative discourse refers to the ability to listen to or read a story, abstract and organize its main points and relevant details, infer information not explicitly stated, and construct an interpretation of the material (Snyder & Downey, 1991). Although studies have documented weak narrative discourse skills in reading-disabled children, the exact nature of the relationship of this higher order processing skill to actual reading performance is unclear (Snyder & Downey, 1991).

Syntactic development and world knowledge relate to reading fluency and comprehension (Snyder & Downey, 1991). Sawyer and Butler (1991) cite evidence that oral syntactic competence accounted for about 50% of second grade reading comprehension.

In addition to being closely intertwined with each other, the phonological, syntactical, and semantic aspects of reading are intimately related to short-and long-term memory, which is critical to automaticity in fluent reading (Sawyer & Butler, 1991).

III. THEORIES AND MODELS OF READING DEVELOPMENT

Neurolinguistic theories and models of reading development have been limited to explaining word-recognition processes. The most prominent of these theories have been the dual-process or dual-route theories (Frith, 1985; Seymour, 1986), which have served as the basis for the classic neurolinguistic model described earlier in this chapter. More recently, connectionist models of the development of word recognition and naming have emerged in the cognitive neuropsychology literature. Both types of models are discussed.

A. Dual-Process Theory

Dual-process theory holds that there are two independent processes operating in the reading and spelling of single words (Pennington, Lefly, Van Orden, Bookman, & Smith, 1987). The indirect route involves phonological processing by grapheme–phoneme correspondence rules (the impaired process in developmental phonological dyslexia), whereas the faster, direct route accesses word-specific knowledge from orthography (the impaired process in developmental surface dyslexia). Frith (cited in Pennington, 1991) postulates three broad stages of reading and spelling development. The first, logographic stage occurs at approximately 4 to 5 yr of age when children develop a sight word vocabulary but no appreciation for the sequential alphabetic code. Children aged 6–7 yr, after 1–1.5 yr of reading instruction, enter the second, alphabetic stage during which they become competent at phonological coding. The onset of the final stage, orthographic processing, occurs around 7 yr and involves going directly from print to meaning.

Recent empirical research has challenged dual-process theories and models. Pennington et al. (1987) tested four predictions of this theory in samples of dyslexic and nondyslexic readers (aged 12 yr to adult) from the same families. The four predictions were (1) that phonological coding skill develops early in normal readers and soon plateaus, whereas orthographic coding skill has a protracted course of development; (2) that adult reading and spelling performance correlates much less with phonological coding skill than with orthographic coding skill; (3) that dyslexics primarily deficient in phonological coding skill should be able to bypass this deficit and eventually close the gap in reading and spelling performance; and (4) that greater phonological coding differences between dyslexic and normal readers would be observed early rather than late in development. None of the four predictions were upheld. Rather, phonological coding skill accounted for 32–53% of the variance in the reading and spelling performance of adult dyslexics, whereas orthographic coding skill accounted for a statistically unreliable portion of the variance. The dyslexics differed little across age in phonological coding skill, and although they surpassed spelling-age controls in orthographic coding skill by adulthood, they did not close the gap in their reading and spelling performance. Finally, the dyslexic individuals had significantly worse phonological coding skill only in adulthood. Similarly, Stuart and Coltheart (1988) present data from their longitudinal study of reading acquisition to dispute these stage theories. Specifically, they found that phonological skills may play a role in the earliest phase of reading acquisition in phonologically adept children, who do not necessarily pass through a logographic stage.

Dual-process models are also criticized for being static box-and-arrow flow diagrams that fail to account for the complex, dynamic, recursive interactions between subcomponents of reading performance (Mozer & Behrmann, 1990) and that certainly do not adequately explain reading acquisition (Holmes-Bernstein & Waber, 1990).

B. Emergence of Connectionist Models

In response to criticisms of classical dual-process models and as a result of increasingly sophisticated computer technology, connectionist models of the development of word reading have appeared in the cognitive neuropsychology literature over the last 5 yr (Berninger, 1990; Seidenberg & McClelland, 1989). Simply speaking, connectionist models (also referred to as parallel, distributed-processing models) are large networks of simple parallel-computing elements or "units" that carry activation values autonomously computed from the values of neighboring network elements via the weighted connections between them (Smolensky, 1988). Encoding occurs by activation of input units that propagate along connections to output units, where the activation computed from the input is encoded. Between the input and output units may be hidden units that represent neither input nor output.

Seidenberg and McClelland (1989) have advanced one such distributed developmental model of word recognition and naming that accounts not only for transitions from beginning to skilled reading and certain forms of developmental and acquired dyslexia, but other aspects of human word-reading performance as well. Although this model assumes that word reading involves the computation of orthographic, phonological, and semantic codes from visual stimuli, and is probably influenced by contextual factors deriving from syntactic, semantic, and pragmatic constraints, only orthographic and phonological units and the hidden units between them comprise the model. Both phonological and orthographic codes are distributed representations (i.e., patterns of activation distributed over a number of primitive representational units). Processing of information between the units is assumed to be interactive and reciprocal. Specifically, in Seidenberg and McClelland's (1989) model, although there are reciprocal connections between the orthographic and hidden units, activation propagates from the hidden units to the phonological units but not in reverse, suggesting that phonological representations do not influence construction of representations at the orthographic level.

The model espoused by Seidenberg and McClelland (1989) differs from dual-process models of word reading in several important ways. First, this connectionist model describes how a single process (from print to sound) is used to read regular words, irregular words, and nonwords in normal readers, whereas a second, indirect route from print to meaning to pronunciation may exist in phonological dyslexia (see also, Seidenberg & McClelland, 1990). Additional ways Seidenberg and McClelland's model contrasts with traditional dual-process models of word reading are that there are no (1) stored lexicons listing pronunciations of all known words, (2) lexical nodes representing individual words, or (3) stored representations of spelling and sound correspondences. Rather, the primary mechanisms involved in lexical processing is activation and spread of activation with spellings and pronunciations of words represented as patterns of activation across output nodes versus access to stored information.

The appeal of parallel, distributed models of word reading over more classic neurolinguistic models is that they are more consonant with Luria's theory of multicomponential, interactive, functional systems in their attempts to account for the reciprocal interaction between phonological skills and reading experience and reading acquisition versus skilled performance (Seidenberg & McClelland, 1989; Stuart & Coltheart, 1988). Furthermore, according to parallel, distributed-processing models of word reading, reading disability might stem from failure of one or more multiple codes to develop, from failure of corresponding codes to become synaptically connected, or from failure of multiple code connections to function temporally in concert (Berninger, 1990; Pennington et al., 1987). Although classical neurolinguistic models also suggest the first two mechanisms of reading dysfunction, they are unable to account for the possibility of temporal dyscoordination among elements in the functional system in any meaningful way.

Seidenberg and McClelland (1989) discuss several important limitations of their model of word reading. First, the model is only concerned with monosyllabic words. Second, these authors have not implemented a process that yields an articulatory-motor response based on computed phonological codes. Finally, issues related to meaning are not addressed. Although possibly superior to existing neurolinguistic models, connectionist models in general also remain limited in their delineation of the neural bases of language and reading because of the lack of information on the dynamic behavior of the nervous system (Smolensky, 1988). Like their predecessors, these models fail to account for the possible role of attentional, frontal, and subcortical functions in language and reading performance and acquisition.

IV. RELATIONSHIPS BETWEEN DEVELOPMENTAL SPEECH–LANGUAGE DISORDERS AND DEVELOPMENTAL DYSLEXIA

Developmental speech–language disorder is commonly defined as a significant speech and/or language deficit despite normal nonverbal intelligence, adequate auditory acuity, and the absence of gross neurological disability (Leonard, Sabbadini, Leonard, & Volterra, 1987), and occurs in 5–15% of children. Developmental dyslexia refers to a difficulty in learning to read and/or spell that is not due to inadequate schooling, peripheral sensory handicaps, acquired brain damage, or low overall IQ and occurs in 5–10% of children (Pennington, 1991).

Although visual deficits and auditory-perceptual deficits seem operative in a minority, the preponderance of empirical evidence suggests significant linguistic deficits, including impaired phonological awareness, short- and, possibly, long-term verbal memory deficits that are probably secondary to basic phonological impairment (Pennington, 1991), naming and word-finding deficits, misarticulations, difficulty repeating nonwords or multisyllabic utterances, and slow vocabulary growth (Snowling, 1991). Precursors of dyslexia observed clinically in some preschoolers include mild speech delay, articulation difficulties, problems learning letter or color names, word-finding problems, missequencing syllables, and problems remembering verbal sequences like addresses, phone numbers, and complex directions (Pennington, 1991).

Because of the strong association between linguistic functioning and reading failure, several researchers have proposed that developmental speech–language impairment and developmental dyslexia are manifestations of the same underlying deficit at different points in development (Benton, 1975; Tallal et al., 1988). This line of research has spawned a growing body of longitudinal investigations of the relationships between developmental speech–language disorders and developmental dyslexia that are relevant to a discussion of the neuropsychology of language and reading development. Although fraught with methodological problems, a brief review of several recent, representative studies will follow.

A. Longitudinal Investigations of Developmental Speech–Language Disorders and Developmental Dyslexia

Tallal and Curtiss (1988) studied 100 specific language-impaired children and 60 matched controls from 4 through 8 yr. They found that expressively impaired preschoolers had better language outcomes at subsequent ages than receptively impaired children. In addition, discriminant function analysis of reading and spelling at age 7 correctly discriminated specific language-impaired children from normal controls with 85% accuracy. Moreover, language-impaired children who no longer had language deficits at 7 yr were dyslexic along with those normal controls who had failed a nonverbal auditory processing task at 4 yr.

Catts (1991) compared kindergarten performance on a battery of standardized language measures and phonological-processing tasks with reading ability in first grade. He found that both receptive and expressive language deficits were related to reading outcome and that children with semantic–syntactic deficits had more reading difficulties than those with primarily articulation problems, a finding that has been replicated by other studies (Bishop & Adams, 1990).

Scarborough (1990) found that 30 month olds with decreased sentence length, syntactic complexity, and pronunciation accuracy on natural language measures were identified as dyslexic in second grade. At 3–3.5 yr, receptive vocabulary and object naming deficits were noted. As other studies have indicated, at 5 yr, subjects evidenced reduced object naming, phonemic awareness, and letter–sound knowledge.

Despite a strong association between preschool speech–language deficits and reading impairment at school age, there is wide-ranging variability in outcomes of preschool speech–language disorders, with 28–75% of speech–language impaired preschoolers exhibiting residual speech–language problems at school age and 52–95% presenting with specific reading impairments (Catts, 1991; Scarborough & Dobrich, 1990). In addition, Scarborough and Dobrich (1990) report on four subjects with broad, severe syntactic, phonological, and lexical deficits at 30 months who approached normal speech–language proficiency by 5 yr. Three of these four children who had appeared to "catch up" in speech–language functioning were identified as severely dyslexic at age 8. These findings suggest the possibility that the functional systems of language and reading are partially intertwined but also somewhat dissociable.

B. Subtypes of Developmental Dyslexia and Developmental Speech–Language Disorders

One of the inherent difficulties in explicating relationships between developmental language disorders and developmental dyslexia is the lack of a standard nosology for either disorder.

Subtyping schemes of developmental dyslexia have ranged from empirical, psychometric approaches popularized by Rourke and his colleagues to more clinical and theoretical approaches (Manis, Szeszulski, Holt, & Graves, 1988; Snowling, 1991). This discussion will highlight clinical and theoretical subtypes that have appeared repeatedly in the literature. For excellent reviews of empirical subtyping efforts and related issues, see Hynd, Connor, and Nieves (1987) and Satz, Morris, and Fletcher (1985).

Based on dual-route theory, analogies drawn between reading and spelling difficulties in developmental dyslexia and acquired dyslexia in adults propose that children who avoid use of phonology while reading are similar to adults with phonological dyslexia, whereas those who rely too heavily on phonology are likened to adults with surface dyslexia (Snowling & Hulme, 1989). These authors comment that fewer references to developmental analogs of deep dyslexia have appeared in the literature. According to such analogies, developmental phonological dyslexics read words better than nonwords and their reading and spelling errors are predominantly dysphonetic, whereas developmental surface dyslexia is characterized by difficulty reading irregular words due to overreliance on phonology, with reading and spelling errors predominantly phonetic in nature (Snowling, 1991).

Two of the more popular clinical–theoretical subtyping schemes in the literature are those of Mattis, French, and Rapin (1975) and Boder (1973). Mattis et al.'s (1975) language disorder syndrome—typified by speech–sound discrimination, verbal comprehension, naming, verbal repetition, and, less frequently, speech–sound sequencing deficits—and Boder's (1973) dysphonetic subtype—characterized by sound–symbol integration deficits resulting in phonetically inaccurate misspellings and misreadings—have marked similarities to developmental phonological dyslexia. On the other hand, Boder's (1973) dyseidetic subtype—characterized by visual-perceptual and visual-memory deficits for letters and whole-word configurations, resultant difficulty in acquiring sight vocabulary, and phonetically accurate misreadings and misspellings—and Mattis et al.'s (1975) visual-spatial perceptual disorder—typified by visual-perceptual and visual-memory deficits for nonverbal material—seem analogous to developmental surface dyslexia. In this respect, adult models of reading performance provide a conceptual starting point for understanding the developmental dyslexias.

In contrast to the language disorder and visual-spatial perceptual syndromes, Mattis et al.'s (1975) third subtype, articulatory and graphomotor dyscoordination syndrome, does not map onto existing adult models. The cardinal symptoms of this syndrome include sound blending difficulties in expressive speech, poor writing and spelling skills, and motor incoordination. In adults with language problems, these types of oral and written language problems are associated with anterior dysfunction (Huntzinger & Harrison, 1992). Furthermore, in an effort to examine the role of prefrontal functions in developmental dyslexia, Kelly, Best, and Kirk (1989) investigated performance on tasks that supposedly measure prefrontal and posterior brain functions in 12-yr-old reading-disabled and non-reading-disabled boys. Results suggested that the prefrontal measures (i.e., Verbal Fluency, Wisconsin Card Sorting Test, and Stroop Color-Word Interference Test) were superior to posterior measures (i.e., Finger Localization, Test of Facial Recognition, Reversals Frequency Test-Recognition subtest, and Boston Naming Test) in discriminating between reading-disabled and non-reading-disabled boys and correctly identified 77% of the boys with reading disabilities from those without reading problems. Depending on additional future validation of this subtype, the articulatory and graphomotor dyscoordination syndrome implies a role for frontal lobe dysfunction in some types of developmental dyslexia.

Formulation of nosologies of developmental speech–language impairments lags behind similar efforts in developmental dyslexia. Besides broad distinctions between developmental language disorders and developmental articulation disorders and between developmental receptive language disorders and developmental expressive language disorders described in the *DSM-III-R* (American Psychiatric Association, 1987), a nosology of clinical language disorder subtypes has been developed by Allen (1989) and her colleagues. This system describes five major constellations—expressive disorders with good comprehension (subdivided into developmental verbal apraxia and phonologic production deficit disorder), verbal-auditory agnosia, phonologic-syntactic disorder, and semantic-pragmatic disorder—and are discussed in greater detail by Allen (1989). Such efforts are critical to advancing understanding of the relationship between developmental speech–language disorders and developmental dyslexia and to delineating neurological systems involved in language and reading acquisition.

V. NEUROANATOMICAL AND ELECTROPHYSIOLOGICAL DATA

Although characterized by methodological inconsistencies and deficiencies, postmortem cytoarchitectonic studies, neuroradiological computed tomography/magnetic resonance imaging (CT/MRI), regional cerebral blood flow (rCBF) studies, and electrophysiological studies provide more direct evidence of structural–functional deviations in the brain and behavioral correlates of dyslexia. These studies are an important source of information in the quest for the neural basis of language and reading (Hynd & Semrud-Clikeman, 1989; Hynd & Willis, 1988).

Postmortem cytoarchitectonic studies document atypical symmetry of the plana temporale, suggest thalamic involvement in reading, and chart widely distributed ectopias and dysplasias that cluster in the left frontal and classical perisylvian language regions, and, to a lesser degree, in the right frontal region in dyslexic brains (Galaburda, Sherman, Rosen, Aboitz, & Geschwind, 1985; Hynd & Semrud-Clikeman, 1989). These abnormalities are proposed to occur in the second trimester during the wave of cell migration that forms the upper cortical layers (Flowers, Wood, & Naylor, 1991). Similarly, a recent postmortem neuropathologic study of a 7-yr-old white female with developmental dysphasia (expressive greater than receptive) revealed atypical symmetry of the plana temporale and gyral displacement along the inferior surface of the Sylvian fissure (Cohen, Campbell, & Yaghmai, 1989, cited in Huettner, 1989).

CT/MRI studies also suggest increased symmetry in the plana temporale and parietal–occipital regions (Hynd & Semrud-Clikeman, 1989). Subsequent MRI studies involving developmentally dyslexic children also found symmetry or reversed asymmetry in the region of the plana temporale (Hynd, Semrud-Clikeman, Lorys, Novey, & Eliopulos, 1990; Jernigan, Hesselink, Sowell, & Tallal, 1991). Additionally, Hynd et al. (1990) found their dyslexic sample to have smaller insular regions bilaterally, and Jernigan et al. (1991) found inferoanterior and superoposterior cerebral asymmetries in their language-learning-impaired sample compared to normal controls, as well as possible additional volume reductions in cortical and subcortical structures. Hynd et al. (1990) also included an attention deficit–hyperactivity disorder (ADHD) sample in their investigation and found that both the ADHD and dyslexic samples had significantly smaller right anterior-width measurements than normal subjects, and that 70% of the normal and ADHD children had the expected left greater than right pattern of plana asymmetry, whereas only 10% of dyslexic children did, suggesting that dyslexia and ADHD are unique and separable syndromes.

Pennington (1991) observes that investigations using electroencephalograms (EEG) and evoked potentials are fairly consistent in demonstrating differences in left hemisphere functioning in dyslexics on tasks that do not involve reading. Badian, McAnulty, Duffy, and Als (1990) examined neuropsychological and electrophysiological performance of 163 boys from kindergarten through fourth grade in an effort to identify possible precursors of dyslexia. All 163 subjects underwent brain electrical activity mapping (BEAM) at age 6 yr, and eight were identified as dyslexic. Preliminary electrophysiological findings suggest a left hemisphere difference, mainly left parietal and frontal, on visual evoked response, and a smaller right hemisphere difference, mainly occipital and parietal, on auditory evoked response. Previously, Duffy, Denckla, Bartels, and Sandini (1980) and Duffy, Denckla, Bartels, Sandini, and Kiessling (1980) found no differences between normal and dyslexic boys during rest, however, found significant differences during listening and reading tasks in Broca's area, the left temporal region and a rough equivalent of Wernicke's area, and the angular gyrus.

Based on the work of Neville and her colleagues (cited in Harter, Anllo-Vento, & Wood, 1989) that documented enhanced visual functioning on event-related potentials in deaf subjects possibly due to functional reorganization of the cortex secondary to their profound auditory deficit, Harter et al. (1989) used event-related potentials to investigate the hypotheses that reading-disabled boys (aged 8–12 yr) would have enhanced spatial attention compared to normal controls and that reading-disabled boys have a neurocognitive processing deficit, as reflected by nonspatial target selection, in this study. The hypothesis that reading-disabled boys would have greater spatial selectivity was confirmed, and the authors concluded that the posterior visual cortices of dyslexics may be more responsive to relevant spatial information, but the sources of this neural difference seemed widely distributed in both anterior and posterior brain regions. Similarly, the hypothesis that the read-ing-disabled boys would have reduced nonspatial target selection was also confirmed, especially in the left occipital lobe. Harter et al. (1989) concluded that their enhanced spatial attention might reflect compensatory neural changes associated with the tendency of reading-disabled individuals to avoid using deficient auditory processes and seeking out experiences that capitalize on their spared visual processes.

Molfese's electrophysiological studies with infants provide further support for the neural basis of language. In 1975, he and his colleagues found evidence for similar stimulus-dependent asymmetries in infants from 1–10 months of age as in adults. Specifically, these researchers found that the auditory evoked response to speech sounds was larger over the left temporal region than the right, while the evoked response to mechanical sounds was larger over the right temporal lobe than the left (Molfese, Freeman, & Palermo, 1975). More recent research of Molfese and his associates has focused on using auditory evoked responses in infancy to predict later language development. Molfese (1992) summarizes these investigations and concludes that brain responses of neonates are useful in making fine discriminations between subjects' perfor-mance on language measures at 4 yr of age. Molfese (1989) also found that auditory evoked responses discriminated between words 14-month-old sub-jects purportedly knew (per maternal report) and those words they did not understand, providing the first indication that auditory evoked responses could be used to detect differences in word meanings in infants.

VI. LANGUAGE AND READING DEVELOPMENT IN EARLY FOCAL BRAIN INJURY

Recent investigations of the development of linguistic functions in children who incurred focal brain injury before 6 months of age have appeared in the literature. This line of research is exciting because it has the potential to provide the kind of developmental information that is necessary for building more realistic, dynamic, and valid models of the neural organization of lan-guage as well as of the brain's capacity for reorganization after injury at a young age (Thal et al., 1991). Some of this work is discussed.

Marchman, Miller, and Bates (1991) investigated phonological analyses of babbling and first words and parental reports of communicative gesture use, word comprehension, and word production in five infants aged 9–22 months who suffered a focal brain injury in the pre- or perinatal period. They found that all five subjects had significant delays in the use of communicative ges-tures, lexical production, and initially produced fewer consonants while bab-bling, although three subjects had phonologically normal vocalizations and had begun to acquire a stable expressive vocabulary by 22 months of age. Marchman et al. (1991) observed that development of consonant production, thought to be a useful index of a child's orientation to the structure of meaningful speech, was clearly related to progress in lexical production while the relationship between communicative gesture use and word production was not as straightforward. Early focal brain insult did not necessarily lead to delayed receptive linguistic abilities in this sample. In this sample, children with anterior lesions evidenced the most improvement in linguistic abilities across the age range studied, whereas the children with occipital or temporal–parietal–occipital lesions continued to exhibit delays at 22 months of age.

Thal et al. (1991) longitudinally investigated lexical development in 27 12–35-month-old children who had incurred focal brain injury prenatally or prior to 6 months of age. They found clear evidence for early delays in both word comprehension and production in the entire sample as well as more comprehension-production dissociations than occur in unimpaired children, with their subjects displaying either unusually high or unusually low lexical comprehension for their level of expressive language development. After expressive vocabulary reached 50 words, they also noted delayed verb production and overreliance on "holistic" formulaic speech (defined as reproduction of relatively long and "underanalyzed" linguistic strings possibly involving some form of rote memory function) as indicated by an atypically high proportion of morphological inflections (e.g., possessive "s" or past tense "-ed") and grammatical function words (e.g., pronouns, prepositions, articles), also known as "closed class" words, relative to the number of nouns and verbs in their vocabularies, thus creating the illusion of advanced grammatical development. The authors contrast such "holistic" linguistic processes with "analytic processes" that involve dissecting linguistic input into its component parts. Based on Bates, Bretherton, and Snyder (1988), who found that children who produced a high number of grammatical function words at 20 months had less productive control over these forms at 28 months than seemingly less grammatically precocious 20-month-old subjects, production of a high proportion of closed class words at 20 months likely reflects atypical linguistic development compared to normally developing children, rather than advanced grammatical development.

Reasoning that disconfirmation of adult-based hypotheses between site of lesion and specific language impairment would advance progress toward more dynamic, developmental models of the brain and language, another purpose of the prospective longitudinal study of Thal et al. (1991) was to examine stable and shifting patterns of associations between site of lesion and specific linguistic deficits. No meaningful differences in linguistic functions based on site of lesion were apparent until 17 months of age. With respect to right–left hemispheric differences, children with right-sided lesions were slightly weaker in verbal comprehension than those with left-sided lesions, which is consistent with adult literature that suggests a role for the right hemisphere in language comprehension (Baynes, 1990). However, anterior–posterior differences in linguistic functions in this young sample differ substantially from the classic brain–language relationships drawn from the adult aphasia literature. Specifically, Thal et al. (1991) found no significant differences in receptive vocabulary between the anterior and posterior lesion groups, contrary to Wernicke's hypothesis. Similarly, in stark contrast to language functions associated with Broca's aphasia in adults, these researchers found that children with left posterior lesions had the greatest expressive language delays and produced fewer grammatical function words, suggesting that left posterior regions, as opposed to anterior regions, mediate acquisition of expressive language and grammatical forms in early language development. Along these same lines, Thal et al. (1991) found that early left posterior lesions were associated with more protracted delays than anterior lesions. Thal et al. (1991) suggest that expressive language delays in children with left posterior damage may derive from impaired sensory analysis required for precise sensorimotor mapping rather than from motor problems as in Broca's aphasia in adulthood. They further speculate that once this phase of language acquisition ends, the role of the left posterior cortex in expressive language may decline.

Children with right hemisphere damage produced a significantly higher proportion of grammatical function words and other closed class morphemes than those with left hemisphere damage, who were essentially within normal limits. Reasoning by analogy to the work of Stiles (1992) on development of visual-spatial pattern analysis, Thal et al. (1991) proposed that the left hemisphere plays a special role in the acquisition of unitary, unanalyzed use of grammatical function words because of left hemisphere specialization for extraction of isolated details versus right hemisphere specialization for comprehension of an overall linguistic configuration or framework. Specifically, they reason that children who produce a high proportion of pronouns and function words in their first word combinations may do so because they have extracted a higher than normal proportion of isolated details from linguistic input. This results in rote production of these little words and possible delay in long-range mastery of grammatical rules, because of the failure to integrate the details into a larger semantic–grammatical framework. Interestingly, along these same lines, based on his findings that adults with right hemisphere damage produce individual ideas and gestures that make sense but are not well integrated by transitions, David McNeill (cited in Adler, 1992b), suggests a similar role for the right hemisphere in providing an overarching, gestalt structure of how single words, sentences, and gestures are interrelated in discourse.

Aram, Gillespie, and Yamashita (1990) investigated reading skills in 30 subjects aged 6–20 yr, who had incurred focal brain lesions at some point in their lives. Twenty subjects had left hemisphere lesions and 10 had right hemisphere lesions. As with Thal et al.'s (1991) longitudinal study of language skills in infants with early focal brain lesions, the results were not entirely consistent with predictions derived from the classic, adult neurolinguistic models. Although subjects with left-hemisphere lesions did more poorly on measures of inferential comprehension than did those with right-sided lesions, there were no significant differences between left- and right-lesioned subjects on phonemic tasks. Similarly, right-lesioned subjects did more poorly on the sound blending task than left-lesioned subjects did, in stark contrast to the large body of evidence for left hemisphere specialization for phonemic tasks. Aram et al. (1990) suggest that the sound blending task may have represented a novel problem-solving activity, which some consider to be a function of right hemisphere processes.

Although Aram et al. (1990) found that brain-injured subjects as a group performed consistently below normal controls on reading-related measures, few of these differences reached statistical significance, suggesting that the majority of individuals with childhood brain injuries learn to read adequately and likely experience considerable functional reorganization of cognitive abilities as they recover. However, the authors comment on the wide individual variability that is obscured by the group findings. Specifically, Aram et al. (1990) found that 25% of left-lesioned subjects, 20% of right-lesioned subjects, and 3.3% of the normal controls had clinically significant reading and spelling problems. Closer analysis of the performance of the reading-impaired subjects revealed no clear relationship between site of lesion and reading deficits in the right hemisphere group. In contrast, although cortical site of lesion did not differentiate between left-lesioned subjects with and without reading disabilities, it is noteworthy that all of the subjects with left hemisphere lesions that involved specific subcortical structures (i.e., head of caudate, putamen, external and internal capsule) presented with reading disorders and associated language and/or memory deficits. Aram et al. (1990) suggest that these subcortical areas may play a vital role in the acquisition of higher cognitive functions that subserve reading development and that these subcortical structures may be less able than cortical areas to reorganize following early injury.

VII. CONCLUSION

The quest for knowledge about the neural bases of language and reading acquisition is necessarily a multidisciplinary endeavor that involves continued neuroanatomical, neuroradiological, electrophysiological, neurolinguistic, neurocognitive, and neurodevelopmental research in normal, speech–language impaired, developmentally dyslexic, and brain-injured populations from infancy through adulthood. The central purpose of this chapter has been to present relevant research from the last 5 yr in each of these disciplines in order to document what is currently known about the neural basis of language and reading acquisition and what work remains to be done.

Based on the current review, recent MRI studies have replicated the adult finding of symmetry or reversed asymmetry of the plana temporale in children with developmental dyslexia (Hynd et al., 1990; Jernigan et al., 1991). However, investigations of language and reading development following early focal brain injury (Aram et al., 1990; Thal et al., 1991) empirically confirm earlier criticisms that the classic neurolinguistic models of language and reading performance derived from the adult literature do not adequately account for language and reading development, and that different models are needed. In this vein, dual-process models of reading are being seriously questioned, and competing parallel, distributed-processing models, which may better account for the complex, dynamic, recursive interactions between subprocesses of reading performance, are being developed. Although parallel, distributed-processing models are more consonant with Luria's notion of interactive functional systems and hold much promise, these models are not yet well developed enough to account for the role of attentional, frontal, and subcortical systems in language and reading performance and acquisition.

In addition to continued and detailed neurolinguistic explication of normal speech–language and reading acquisition, along with development of age-appropriate measures to assess various neurolinguistic processes across development, more prospective, longitudinal investigations of the course of developmental speech–language disorders and developmental dyslexia are needed to expand knowledge related to the neural bases of language and reading development. Development of a standard nosology for identifying developmental speech–language disorders and developmental dyslexias is critical to the utility of such longitudinal investigations.

Given the multifaceted, multidisciplinary nature of the quest for the neural bases of language and reading development, a nation-wide, prospective, longitudinal population survey from infancy through adulthood would go a long way toward advancing knowledge in this area and integrating relevant research findings from diverse fields. Such a project would necessarily include the development of age-appropriate measures of language, reading, attentional, and frontal processes at various developmental levels, involve neuroradiological and electrophysiological measurements, and attend to relevant environmental and experiential factors (e.g., socioeconomic status, otitis media, instructional programs) and emotional factors. Based on recent research findings (Gorman, 1992), a project of this type, or any future research in this area, would do well to attend to possible gender differences in functional organization of language and reading processes as well.

REFERENCES

Adler, T. (1992a, April). Babies screen out meaningless sounds. *APA Monitor,* 44.
Adler, T. (1992b, April). Gestures offer clues on speech and brain. *APA Monitor,* 41.
Allen, D. A. (1989). Developmental language disorders in preschool children: Clinical subtypes and syndromes. *School Psychology Review, 18,* 442–451.

American Psychiatric Association. (1987). *Diagnostic and statistical manual of mental disorders (3rd ed., revised)*. Washington, D.C.: Author.

Aram, D. M., Gillespie, L. L., & Yamashita, T. S. (1990). Reading among children with left and right brain lesions. *Developmental Neuropsychology, 6,* 301–317.

Badian, N. A., McAnulty, G. B., Duffy, F. H., & Als, H. (1990). Prediction of dyslexia in kindergarten boys. *Annals of Dyslexia, 40,* 152–169.

Bastian, H. C. (1898). *Aphasia and other speech defects*. London: H. K. Lewis.

Bates, E., Bretherton, I., & Snyder, L. (1988). *From first words to grammar: Individual differences and dissociable mechanisms*. New York: Cambridge University Press.

Bates, E., Thal, D., Whitesell, K., Fenson, L., & Oakes, L. (1989). Integrating language and gesture in infancy. *Developmental Psychology, 25,* 1004–1019.

Baynes, K. (1990). Language and reading in the right hemisphere: Highways or byways of the brain? *Journal of Cognitive Neuroscience, 2,* 159–179.

Benton, A. L. (1975). Developmental dyslexia: Neurological aspects. In J. Friedlander (Ed.), *Advances in Neurology (Vol. 7)*. New York: Raven.

Berninger, V. W. (1990). Multiple orthographic codes: Key to alternative instructional methodologies for developing the orthographic-phonological connections underlying word identification. *School Psychology Review, 19,* 518–533.

Bishop, D. V., & Adams, C. (1990). A prospective study of the relationship between specific language impairment, phonological disorders, and reading retardation. *Journal of Child Psychology & Psychiatry & Allied Disciplines, 31,* 1027–1050.

Blachman, B. (1991). Early intervention for children's learning problems: Clinical applications of the research in phonological awareness. *Topics in Language Disorders, 12,* 51–65.

Boder, E. (1973). Developmental dyslexia: A diagnostic approach based on three atypical reading-spelling patterns. *Developmental Medicine and Child Neurology, 15,* 663–687.

Catts, H. W. (1991). Early identification of dyslexia: Evidence from a follow-up study of speech–language impaired children. *Annals of Dyslexia, 41,* 163–177.

Chall, J. S. (1987). Reading development in adults. *Annals of Dyslexia, 37,* 240–251.

de Hirsch, K. (1984). *Language and the developing child*. Orton Dyslexia Society: Baltimore, MD.

Duffy, F., Denckla, M. B., Bartels, P. H., & Sandini, G. (1980). Regional differences in brain electrical activity by topographic mapping. *Annals of Neurology, 7,* 412–420.

Duffy, F., Denckla, M. B., Bartels, P. H., Sandini, G., & Kiessling, L. A. (1980). Dyslexia: Automated diagnosis by computerized classification of brain electrical activity. *Annals of Neurology, 7,* 421–428.

Eimas, P. D., Siqueland, E. R., Jusczyk, P., & Vigorito, J. (1971). Speech perception in infants. *Science, 171,* 303–306.

Flowers, D. L., Wood, F. B., & Naylor, C. E. (1991). Regional cerebral blood flow correlates of language processes in reading disability. *Archives of Neurology, 48,* 637–643.

Frith, U. (1985). Beneath the surface of developmental dyslexia. In K. E. Patterson, J. C. Marshall, & M. Coltheart (Eds.), *Surface dyslexia*. London: Routledge & Kegan-Paul.

Galaburda, A. M., Sherman, G. F., Rosen, G. D., Aboitz, F., & Geschwind, N. (1985). Developmental dyslexia: Four consecutive patients with cortical anomalies. *Annals of Neurology, 18,* 222–233.

Geschwind, N. (1962). The anatomy of acquired disorders of reading. In J. Money (Ed.), *Reading disability*. Baltimore: John Hopkins University Press.

Geschwind, N. (1972). Language and the brain. *Scientific American, 226,* 76–83.

Goldfield, B. A., & Reznick, J. S. (1990). Early lexical acquisition: Rate, content, and the vocabulary spurt. *Journal of Child Language, 17,* 171–183.

Gorman, C. (1992, January 20). Sizing up the sexes. *Time,* 42–51.

Harter, M. R., Anllo-Vento, L., & Wood, F. B. (1989). Event-related potentials, spatial orienting, and reading disabilities. *Psychophysiology, 26,* 404–421.

Holmes-Bernstein, J., & Waber, D. P. (1990). Developmental neuropsychological assessment:

The systemic approach. In A. A. Boulton, G. B. Baker, & M. Hiscock (Eds.), *Neuromethods*. Clifton, N.J.: Humana Press.

Huettner, M. I. S. (1989). Neurological basis of language and reading. *Learning and Individual Differences, 1,* 407–421.

Huntzinger, R. M., & Harrison, D. W. (1992, February). *Fluent versus nonfluent subtypes of dyslexic readers.* Paper presented at the meeting of the International Neuropsychological Society, San Diego, CA.

Hynd, G. W., Connor, R. T., & Nieves, N. (1987). Learning disabilities subtypes: Perspectives and methodological issues in clinical assessment. In M. G. Tramontana, & S. R. Hooper (Eds.), *Assessment issues in child neuropsychology.* New York: Plenum Publishing Corporation.

Hynd, G. W., & Hynd, C. R. (1984). Dyslexia: Neuroanatomical/neurolinguistic perspectives. *Reading Research Quarterly, 19,* 482–498.

Hynd, G. W., & Semrud-Clikeman, M. (1989). Dyslexia and brain morphology. *Psychological Bulletin, 106,* 447–482.

Hynd, G. W., Semrud-Clikeman, M., Lorys, A. R., Novey, E. S., & Eliopolos, D. (1990). Brain morphology in developmental dyslexia and attention deficit disorder/hyperactivity. *Archives of Neurology, 47,* 919–926.

Hynd, G. W., & Willis, G. W. (1988). *Pediatric neuropsychology.* New York: Grune & Stratton, Inc.

Jernigan, T. L., Hesselink, J. R., Sowell, E., & Tallal, P. A. (1991). Cerebral structure on magnetic resonance imaging in language- and learning-impaired children. *Archives of Neurology, 48,* 539–545.

Kelly, M. S., Best, C. T., & Kirk, U. (1989). Cognitive deficits in reading disabilities: A prefrontal cortical hypothesis. *Brain & Cognition, 11,* 275–293.

Kirk, U. (1983). Introduction: Toward an understanding of the neuropsychology of language, reading and spelling. In U. Kirk (Ed.), *Neuropsychology of language, reading, and spelling.* New York: Academic Press.

Leonard, L. B., Sabbadini, L., Leonard, J. S., & Volterra, V. (1987). Specific language impairment in children: A cross-linguistic study. *Brain & Language, 32,* 233–252.

Manis, F. R., Szeszulski, P. A., Holt, L. K., & Graves, K. (1988). A developmental perspective on dyslexic subtypes. *Annals of Dyslexia, 38,* 139–153.

Marchman, V. A., Miller, R., & Bates, E. A. (1991). Babble and first words in children with focal brain injury. *Applied Psycholinguistics, 12,* 1–22.

Mattis, S., French, J. H., & Rapin, I. (1975). Dyslexia in children and young adults: Three independent neuropsychological syndromes. *Developmental Medicine and Child Neurology, 17,* 150–163.

Mayeux, R., & Kandel, E. R. (1985). Natural language, disorders of language, and other localizable disorders of cognitive functioning. In E. R. Kandel, & J. H. Schwartz (Eds.), *Principles of neural science (2nd ed.).* New York: Elsevier.

Molfese, D. L. (1989). Electrophysiological correlates of word meanings in 14-month-old human infants. *Developmental Neuropsychology, 5,* 79–103.

Molfese, D. L. (1992). The use of auditory evoked responses recorded from newborn infants to predict language skills. In M. G. Tramontana, & S. R. Hooper (Eds.), *Advances in child neuropsychology Vol. 1.* New York: Springer-Verlag, Inc.

Molfese, D. L., Freeman, R. B., & Palermo, D. S. (1975). The ontogeny of brain lateralization for speech and nonspeech stimuli. *Brain & Language, 2,* 356–368.

Mozer, M. C., & Behrman, M. (1990). On the interaction of selective attention and lexical knowledge: A connectionist account of neglect dyslexia. *Journal of Cognitive Neuroscience, 2,* 96–123.

Newcombe, F., Phil, D., and Marshall, J. C. (1981). On psycholinguistic classification of the acquired dyslexias. *Bulletin of the Orton Society, 31,* 29–46.

Pennington, B. F. (1991). *Diagnosing learning disorders: A neuropsychological framework.* New York: Guilford Press.

Pennington, B. F., Lefly, D. L., Van Orden, G. C., Bookman, M. O., & Smith, S. D. (1987). Is phonology bypassed in normal or dyslexic development? *Annals of Dyslexia, 37,* 62–89.

Recht, D. R., Newby, R. F., & Caldwell, J. (1992, February). *Phonological processing patterns in two subtypes of children with dyslexia.* Paper presented at the meeting of the International Neuropsychological Society, San Diego, CA.

Rourke, B. P. (1985). *Neuropsychology of learning disabilities: Essentials of subtype analysis.* New York: Guilford Press.

Ryalls, J., Beland, R., & Joanette, Y. (1990). Contributions of linguistic approaches to human neuropsychology: Aphasia. In A. A. Boulton, G. B. Baker, & M. Hiscock (Eds.), *Neuromethods.* Clifton, N.J.: Humana Press.

Satz, P., Morris, R., & Fletcher, J. M. (1985). Hypotheses, subtypes, and individual differences in dyslexia: Some reflections. In D. B. Gray, & J. F. Kavanagh (Eds.), *Biobehavioral measures of dyslexia.* Parkton, MD: York Press.

Sawyer, D. J., & Butler, K. (1991). Early language intervention: A deterrent to reading disability. *Annals of Dyslexia, 41,* 55–79.

Scarborough, H. S. (1990). Very early language deficits in dyslexic children. *Child Development, 61,* 1728–1743.

Scarborough, H. S., & Dobrich, W. (1990). Development of children with early language delay. *Journal of Speech and Hearing Research, 33,* 70–83.

Seidenberg, M. S., & McClelland, J. L. (1989). A distributed, developmental model of word recognition and naming. *Psychological Review, 96,* 523–568.

Seidenberg, M. S., & McClelland, J. L. (1990). More words but still no lexicon: Reply to Besner et al. (1990). *Psychological Review, 97,* 447–452.

Seymour, P. H. K. (1986). *Cognitive analysis of dyslexia.* London: Routledge & Kegan-Paul.

Smith, C. A., & Sachs, J. (1990). Cognition and the verb lexicon in early lexical development. *Applied Psycholinguistics, 11,* 409–424.

Smolensky, P. (1988). On the proper treatment of connectionism. *Behavioral and Brain Sciences, 11,* 1–74.

Snowling, M. J. (1991). Developmental reading disorders. *Journal of Child Psychology & Psychiatry & Allied Disciplines, 32,* 49–77.

Snowling, M., & Hulme, C. (1989). A longitudinal case study of developmental phonological dyslexia. *Cognitive Neuropsychology, 6,* 379–401.

Snyder, L. S., & Downey, D. M. (1991). The language–reading relationship in normal and reading disabled children. *Journal of Speech and Hearing Research, 34,* 129–140.

Stiles, J. (1992, February). *Spatial cognitive development in children with focal brain injury.* Paper presented at the meeting of the International Neuropsychological Society, San Diego, CA.

Stoel-Gammon, C. (1991). Normal and disordered phonology in 2-year-olds. *Topics in Language Disorders, 11,* 21–32.

Stuart, M., & Coltheart, M. (1988). Does reading develop in a sequence of stages? *Cognition, 30,* 139–181.

Szeszulski, P. A., & Manis, F. R. (1990). An examination of familial resemblance among subgroups of dyslexics. *Annals of Dyslexia, 40,* 180–191.

Tallal, P., & Curtiss, S. (1988, January). *From developmental dysphasia to dyslexia: A neurodevelopmental continuum.* Paper presented at the meeting of the International Neuropsychological Society, New Orleans, LA.

Tallal, P., Curtiss, S., & Kaplan, R. (1988). The San Diego longitudinal study: Evaluating the outcomes of preschool impairment in language development. In S. E. Gerber, & G. T. Mencher (Eds.), *International perspectives on communication disorders* (pp. 86–126). Washington, D.C.: Gallaudet University Press.

Thal, D. J. (1991). Language and cognition in normal and late-talking toddlers. *Topics in Language Disorders, 11,* 33–42.

Thal, D. J., Marchman, V. A., Stiles, J., Aram, D., Trauner, D., Nass, R., & Bates, E. (1991). Early lexical development in children with focal brain injury. *Brain & Language, 40,* 491–527.

Van Lancker, D. (1987). Nonpropositional speech: Neurolinguistic studies. In A. Ellis (Ed.), *Progress in the psychology of language.* Hillsdale, N.J.: Lawrence Erlbaum & Associates.

Zecker, S. G. (1991). The orthographic code: Developmental trends in reading-disabled and normally achieving children. *Annals of Dyslexia, 41.*

3

Intelligence and the Brain

Douglas K. Detterman

I. INTRODUCTION

It seems obvious there is a relationship between human intelligence and the brain. The brain weighs 3–4 lbs. It is composed of 100 billion neurons (Fischbach, 1992). Each neuron is composed of an axon and multiple dendrites. Besides nerve cells there are complex chemical factories generating the neural transmitters and inhibitors that allow nerve cells to talk to each other. The neural environment is further complicated by the effects of hormones and other agents not produced in the brain. Over half of our 10,000 genes are thought to regulate nervous system development and functioning.

Now consider intelligence. There is no doubt there is measurable variability in human ability. Intelligence tests are highly reliable. If two versions of the same test are given within a short span of time, the correlation between the tests will be as high as any psychological measure can be made to be. Intelligence tests predict important things like number of years of schooling completed, how well a person can be expected to do in school, and the score on other tests of intelligence. The reasons for these correlations may be debatable but the empirical fact of the correlations is undisputable. IQ tests correlate with other measures of mental functioning higher than any other kind of psychological test correlates with other measures of its own kind. Even more impressive is the correlation of IQ tests with validity measures. Validity measures for IQ tests, as a group, are higher than for any other kind of psychological tests. For example, personality tests don't correlate nearly as highly with appropriate personality validity indices as IQ tests correlate with appropriate intellectual validity indices. There is little doubt that it is possible to measure intelligence both reliably and validly. The question is why intelligence is such a reliable and valid measure.

One hypothesis suggested for at least a century is that intelligence is partially biological and the source of those biological differences is in the brain. This hypothesis has great intuitive appeal for some and absolutely no appeal for others. Everyone agrees that all differences must ultimately be reflected in biology. Even the most transitory stimulus event must have some biological effect on the organism that notices it. The important question is whether or not any differences in IQ are caused by hard-wired, unchangeable brain differences.

The simplest, most parsimonious hypothesis is that the brain has nothing to do with intelligence. This hypothesis is obviously wrong. Simple observation shows some relationship between brain and intelligence. Although 50% of mental retardation has no identifiable cause, 50% does. When a particular disorder causes mental retardation, there is always some identifiable brain pathology. It may not be possible to explain how the pathology causes mental retardation. Brain pathology is such an invariant feature of mental retardation syndromes that there is little doubt of its importance as a cause of impaired intelligence. Similar observations apply to brain trauma, anoxia, or infection. All can be linked to impaired intellectual functioning.

Gross evidence from mental retardation syndromes and brain trauma show that the brain must somehow be associated with intelligence. They do not tell us how they are linked, though. All that these observations tell us is that impaired brains do not work like unimpaired brains. Perhaps individual differences in intelligence could be accounted for by the amount of damage to a person's brain. This is called the cumulative deficit hypothesis. Information about the brain impairment can only give us clues about the parts of the brain involved in intelligence.

Evidence from damage cannot tell us if differences in the damaged parts of the brain would lead to differences in intelligence if they were not damaged. Finding subtle differences in the brain that cause differences in intelligence will require another kind of evidence. Researchers will have to show differences in normal brain structures associated with differences in IQ. Even this evidence will be only suggestive. To be completely convincing, how the differences in brain structure arise would have to be explained. The most convincing explanation would be one that describes the origin of differences beginning with the gene, proceeds to developmental structural differences caused by genetic influences, and ends by describing behavior as a function of differences in structure. Such an account would have to include environmental effects at every level and stage. Only with such a complete explanation could we be certain of the relationship between brain and intelligence.

A complete explanation of the brain–intelligence relationship will be some time in coming. There is nothing even close to it now. A complex human characteristic like intelligence and a complex biological structure like the brain are not going to converge easily. It is apparent they must eventually merge. Each has a substantial heritage to give the other. Those who study intelligence are expert at precise measurement of cognitive behavior. Neurobiologists, on the other hand are expert in the neurobiological mechanisms and methodologies that might provide the ultimate explanation of intelligence.

How will this merger take place? As the proceeding studies cited show, the confluence will probably be initiated by behavioral scientists. Biologists have little tradition of studying individual differences. What attracts behavioral scientists are the powerful techniques being developed for studying the brain.

The kinds of studies done will be determined by the theories that researchers hold. Those who believe that intelligence (particularly g) is a single thing will look for global characteristics of the nervous system that could explain individual differences in intelligence. For example, Eysenck (1986) and his students, who consider g to be unitary, study such biological measures as electrical conduction speed and complexity in the brain.

My theory (Detterman, 1987) is that intelligence is a complex system of independent parts. Measures of intelligence reflect global system functioning. Because of my theoretical orientation, my prejudice is to expect that several, perhaps many, measures of brain functioning will be related to intelligence. I also expect that no single biological measure will account for a large portion of the variance in intelligence. It is too early to argue that biological evidence supports one theory of intelligence or another, as will be clear from the studies reported in the following sections. However, the theoretical orientation of the investigator should be kept in mind when considering studies relating biology and intelligence. The variables selected for study may represent the investigator's bias as much as anything else.

Are differences in IQ related to brain differences? What seems intuitively obvious is not always easy to support scientifically. The search for exactly what differences in the brain cause differences in IQ has just begun. Powerful new technologies are just now becoming available to behavioral researchers. This chapter is a progress report on the search for the brain–intelligence connection.

II. HEAD INJURY

One obvious way of studying how the brain works is to study how localized trauma to different parts of the brain affects intelligence. Such studies are easily carried out in rats. In Lashley's (1929; see also Beach, Hebb, Morgan, & Nissen, 1960) classic studies, portions of rats' brains were ablated. The maze behavior of the ablated rats was then studied. Lashley concluded his studies by formulating two general principles: mass action and equipotentiality. Mass action meant the brain acted as a whole, and equipotentiality meant that all parts of the brain were equal. In short, Lashley found that lesions to different parts of the brain didn't have differential effects on maze behavior. The only thing that mattered was how big the lesions were. The bigger the lesions, the more they affected maze behavior.

Despite Lashley's experiments, most descriptions of the human brain localize functions to specific areas. Clever simulations by Wood (1978, 1980) show that it is possible to have an integrated, distributed system that produces symptoms of localization even though nothing is localized. Apparent localization results because of the way the system operates. Wood's model explains both Lashley's findings and findings reporting localization of functioning.

Thompson, Crinella, and Yu (1990) did experiments similar to Lashley's but with larger numbers of rats, more sophisticated lesion positioning, and more diverse behavioral tests. Factor analysis of their behavior data from lesioned rats showed that there was a general learning system in the brain. This learning system seemed to be importantly dependent on deeper structures of the brain. Most studies of brain injury assume that the cortex is all-important to intelligence. According to Thompson, Crinella and Yu, studying the cortex may be the wrong approach to studying intelligence.

It is not possible to produce experimental lesions in humans. Wars, however, provide subjects who have limited damage to the cortex. It seems that a gunshot wound in the brain is not as critical as it sounds. Wounds to the cortex, while serious, often allow for nearly full recovery. Do such wounds impair intelligence?

There have been studies of World War II (Weinstein & Teuber, 1957) and Vietnam (Grafman et al., 1988; Grafman, Salazar, Weingartner, Vance, & Amin, 1986) survivors of gunshot wounds to the brain. Since all of the subjects were in the military, they were given a full-length intelligence test before their injury. Retesting them some years after the injury tells if the injury had any impact on intelligence.

Most people, when asked, believe that the kind of damage to your brain that these veterans received would result in a large drop in IQ. Even the most serious injury resulted in an IQ decline of not more than ten points. The study of the Vietnam veterans showed that, for all cases, there was no significant decline in IQ. In a later paper, Grafman et al. (1988) found that while the volume of brain tissue lost was not related to changes in IQ, it was related to declines in tests of specific cognitive abilities. Perhaps the kinds of deficits that occur from these injuries are more subtle than can be detected by gross tests of human intelligence. A battery of more specific cognitive tests might do a better job of identifying the effects bullet wounds have on the cortex. It is surprising the effects of bullet wounds are so subtle. The small effects found suggest the cortex is not critically important in human intelligence. The most central processes in human intelligence may be more closely related to structures deeper in the human brain, as suggested by Thompson, Crinella, and Yu (1990).

III. ANATOMY

The study of anatomical characteristics has been surprisingly unproductive in yielding information about the important processes in human intelligence. There are several problems in relating anatomy to intelligence. First, anatomical areas of the brain have no necessary relationship to human behaviors. Second, areas of the brain show individual differences that are difficult to measure reliably even at autopsy much less in living subjects.

A. Gross Anatomy

Since the brain was first studied systematically, and even before, there have been attempts to find relationships between the gross anatomical features of the brain and human characteristics like intelligence. One notorious effort was known as phrenology. According to the theory of phrenology, bumps on a person's head represented prominences in the brain that directly influenced various human characteristics. The size of a particular bump on the head reflected how much of a particular characteristic the person had. As late as the 1930s, this "science" was seriously practiced. Not only was it practiced, but phrenologists developed the first totally automated test. A machine that had a number of sensors over critical parts of the brain was used for testing. A metal cap like a large beauty shop hair dryer was placed on the testee's head. The sensors in the cap measured the height of critical bumps on the head of the person being tested. Based on the reading of sensors in the cap, a printout detailing the subject's intelligence and personality was given. I had the opportunity to be tested in one of these machines on two separate occasions. The results the machine produced were as reliable as an intelligence test. The validity of such tests has never been shown, however.

B. Anatomical Features

A large number of anatomical features have been localized in the brain. Some of these areas, like Broca's and Wernicke's area have even been associated with specific human cognitive functions that must be important in intelligence. Damage to such areas has also been associated with impairments in specific cognitive functioning. Such impairments are the usual method for attributing functional importance to a particular area of the brain. Despite these kinds of evidence, to my knowledge there has never been any direct evidence that any characteristics of these areas account for differences in human intelligence. It must be remembered that describing the function of an area of the brain does not mean that area of the brain will result in individual differences. A different kind of evidence is necessary to reach that conclusion. Differences in the size or functioning of the area must be associated with psychological individual differences.

It is not that people haven't looked for the kind of differences that could account for differences in intelligence. Einstein's brain was thoroughly examined and no one reported any distinguishing features that would account for his accomplishments in whole or part. It is, of course, possible that we just don't know what to look for. There have, however, been some recent developments that suggest that gross anatomical features of the brain may be more important than previously thought.

1. Brain Size   There has been a long-standing controversy in psychology about the relationship between brain size and intelligence. Various reviews of the issue have concluded that brain size and intelligence correlate between .10 and .35 (e.g., Stott, 1983). There is an impediment to reaching any firm conclusion about this issue. Measuring brain size requires dead subjects, whereas measuring intelligence requires live subjects. Most data were based on dead persons who happened to have intelligence test scores.

New imaging techniques have changed all that. It is now possible to measure brain size in living persons. Magnetic resonance imaging (MRI) produces a complete picture of the brain. The picture provides superlative detail and even distinguishes anatomical regions based on tissue density. Most important, the MRI produces a set of computer data that can be used to produce a picture of the brain or to study its gross anatomical structure. The computer data allow absolute comparison of the size of the brain and even different parts of the brain.

There have been some studies using this technology. Willerman, Schultz, Rutledge, and Bigler (1992) found a correlation of .35 between brain size as assessed by MRI and IQ. Though there are many methodological issues associated with a study like this, clearly the technique will allow better measurement than has been obtained with previous methods.

A good example of a study that distinguishes gross anatomical areas is the study by Yeo, Turkheimer, Raz, and Bigler (1987). They compared the left and right hemisphere sizes with discrepancies in the Wechsler Adult Intelligence Scale (WAIS) verbal and performance scales. Differences between the left minus right hemisphere were moderately correlated with the difference between the verbal and performance scales on the WAIS. The reasoning for this finding is that if verbal abilities are localized in the left hemisphere of the brain, the people who do better on the verbal scales should have larger left hemispheres.

What makes such studies possible is that an MRI or any other imaging technique captures the image as numerical data. These numerical data can be analyzed like any other data. Current technology requires a great deal of human assistance because of the difficulty of recognizing brain structures. Future developments may make automatic recognition of brain parts possible.

2. Sex Differences   A recent controversial topic has been sex differences in gross anatomy. Ankney (1992) and Rushton (1992) reported that men have larger brains than women based on gross anatomical data. This difference is found even when body size and other relevant factors are accounted for. Men's brains are about 100 g larger than women's brains.

Willerman et al. (1992) found that the relative difference between hemispheres predicted different things in men and women. For men, a relatively larger left hemisphere predicted superior verbal performance compared to nonverbal performance. For women, just the opposite was true: a larger left hemisphere resulted in better nonverbal performance. The authors interpret this result as showing that men's and women's brains have functionally different organizations.

An implication, when not explicitly stated, is that the kind of sex differences being found in these studies could explain the well-known differences between males and females in spatial and higher math ability. During the evolution of the human brain, men were hunters. Hunting requires skills of spatial visualization. Women, on the other hand, took care of children, minded the home, and farmed or gathered. All these activities require social and language skills. The additional weight of men's brains, so the argument goes, represents areas associated with spatial skills.

Remember that sex differences have just begun to be investigated using new technologies like MRI. The old methods involving physical measurements of cranial capacity or brain size were often imprecise and difficult to carry out. Now, or at least eventually, it will be possible to get very accurate estimates of brain size and even the relative size of various parts of the brain.

C. Conclusions

Gross anatomical structure gives little evidence that the brain has much to do with intelligence. Damage to the cortex produces only small decrements in intelligence. Measures of gross anatomical features, even when obtained with MRI, have only small to moderate relationships to intelligence. It would be easy to conclude that the brain has little to do with intelligence.

Several issues should be considered before reaching this conclusion. First, it is unlikely that one aspect of brain functioning will account for more than a moderate amount of the variance in intelligence. The brain is a complex system. Many parts of this system are likely to produce differences in intelligence. If there were only ten equally important biological factors that produced differences in intelligence, each would account for 10% of the biological variance in intelligence. Correlations between IQ and these biological factors would all be under .32. (The square of the correlation coefficient is the proportion of variance accounted for: $.32^2 = .10$).

Second, questions that could be asked using gross anatomical information were limited before the recent availability of MRI. Even with MRI some questions will never be answered using gross anatomical information. For example, brain structures deeper than the cortex may be importantly implicated in intelligence. It is not possible to study the effects of damage to these structures on intelligence because the structures are so important to the organism that damage to them causes death. There is little information about these deeper structures and their relationship to intelligence in humans. Some techniques discussed in the following sections may be helpful in understanding if they affect intelligence.

IV. MICROSCOPIC STRUCTURE

Hebb (1949, 1980) proposed that intelligence was adaptability of the central nervous system. A key idea in his theory was the concept of cell assemblies. In simplified terms, cell assemblies were formed when sets of neurons that were stimulated together came to act as a unit. According to Hebb, intelligence could best be understood by studying the function of neurons.

Hebb's central thesis is supported by the study of the effects of differential rearing on brain development. Rats reared in an enriched environment show brain changes and improved learning (e.g., Rosenzweig & Bennett, 1977). Brain changes interact with experiences during critical times of development (Hubel & Weisel, 1970).

As Hebb suggested, brain changes associated with intelligent behavior like maze learning are present at the neuronal level. Studies like the ones cited above often find evidence of brain changes, including dendritic changes. The number of dendrites (the multiple branches of the nerve that receive signals from other nerves' axons and carry the signal toward the cell body) in the nervous system is important for efficient function.

Dendritic Sprouting and Pruning

An excess of dendrites sprouts early in development (Huttenlocher, 1979; Purpura, 1975). Maximal dendritic sprouting in human infants occurs at age 1 yr (Huttenlocher, 1984; Huttenlocher, deCourten, Garey, & Van der Loos, 1982). Pruning of excess dendrites follows. The dendrites pruned may be those that receive the least stimulation. This would provide a mechanism for the functional shaping of the brain by experience (Green & Greenough, 1986; Green, Greenough, & Schlumpf, 1983; Greenough, McDonald, Parnisari, & Camel, 1986).

Though the mechanism of dendritic pruning is an interesting one, there is no evidence that it contributes to individual differences in the biological basis of intelligence. There is no doubt that it could provide a mechanism to account for how experience could raise or lower intelligence. There is no evidence, however, that people vary in their ability to be efficient dendritic pruners. This is an important distinction. Showing large group effects of dendritic pruning on things like maze learning by varying early experience is possible. These studies do not show individual differences, though. It could very well be that all brains work alike when it comes to dendritic pruning; however, if all brains worked alike, there would be no biological basis for explaining differences in intelligence.

Humans are alike in their dendritic pruning according to a study by Huttenlocher (1974). He examined mentally retarded persons' brains for differences in neural pruning. No differences were found for older mentally retarded persons, though it had been shown that younger mentally retarded persons showed lower levels of pruning than age-matched children of normal intelligence. This finding suggested that mentally retarded persons were delayed in neural pruning, not qualitatively different. In short, the finding suggests that neural pruning probably could not explain the mental retardation or the biological basis for the differences in intelligence.

V. ELECTROPHYSIOLOGICAL METHODS

Electrophysiological measures of brain functioning are usually indirect measures. In humans, they are taken from the surface of the body even when the goal is to get measures of electrical activity within the body. For measures of brain activity like electroencephalograms (EEG) and average evoked potentials (AEP, also called event-related potentials, ERP), measures are taken from the surface of the head. In the techniques most widely used it is not possible to identify where the electrical signals being recorded originate. Placing an electrode over the occipital lobe does not mean that the electrical activity being recorded comes from the occipital lobe, although that is what is usually assumed.

Another problem with electrical measures of nervous system activity is that they are subject to artifact. For example, the recording of AEPs involves very small currents in comparison to the electrical activity associated with eye movements. Without controlling for or eliminating trials where eye movements occur, the AEP data will be overwhelmed by noise from eye movements. Even if eye movements are controlled, a single AEP trial contains a lot of electrical noise. Only by repeating a trial many times is noise averaged out over trials.

Complexity is a problem in almost all measures of the intact human nervous system, but it is a particularly acute problem with electrophysiological measures. Electrophysiological measurements require collection of a huge amount of data. These data are unusable in raw form. They must be cleaned and then subjected to data reduction methods. Sample sizes are usually small. It is no wonder there are so many methodological debates.

Despite these difficulties, electrophysiological methods have several advantages. Of all methods available, electrophysiological methods are the only ones that allow the real time observation of brain activity. Newer techniques being developed may be able to localize the source of electrical activity within the brain. Theoretically, it should be possible to trace a stimulus–response action chain through the brain, identifying those parts of the brain most active during different parts of the activity being performed.

Another advantage of electrophysiological methods is that they are totally noninvasive and inexpensive to apply. For example, positron emission tomography (PET) scans require a particle accelerator to produce atomic isotopes, a several million-dollar investment beyond the budgets of average psychology laboratories. Even if atomic accelerators become cheap enough for everyone to own, the use of atomic isotopes is more complex than giving paired-associate lists.

A. Neurometrics

There is absolutely no doubt that electrophysiological measures can and do provide the same kinds of information that intelligence tests provide. John and his colleagues (John, 1977; John et al., 1977) have shown that it is possible to make the same decisions made with IQ tests basing those decisions on AEP and EEG measures. Their predictions of learning disability and mental retardation using AEP and EEG are as accurate as traditional IQ tests.

In their procedure, each subject is given a standardized series of electrophysiological measures. These measures are then reduced and compared to norms allowing classification of the subject. A machine is commercially available and is reported to cost about $80,000. This is about 200 times more expensive than the equipment needed for a standard intelligence test. The only reason to prefer the more expensive method developed by John would be if it gave additional information useful in understanding the source of mental retardation or learning disability. To my knowledge, it does not provide any additional information beyond classification.

What the work by John and his colleagues *does* show is that electrophysiological measures and IQ tests contain the same sources of variance. Electrophysiological measures can predict IQ tests as well as IQ tests predict each other (at least for disabled persons). This is not a trivial point. It suggests that if we understood the electrical activity of the brain, we would understand intelligence.

B. Electroencephalographic Measures

The EEG is the oldest of the electrophysiological measures used to study brain activity's relation to intelligence. EEGs are obtained by attaching 12 or more electrodes to the brain and measuring gross electrical activity. Electrical activity is then analyzed by regularity, frequency, and amplitude. Characteristic waveforms, of which there are at least seven, are named with Greek letters. These characteristic waveforms are associated with different kinds of activity. For example, alpha waves are typical of a relaxed, awake state and have a high amplitude and frequencies of 8–13 Hz.

One question that has frequently been raised is the reliability of EEG measures. Some think EEGs are no more valid than phrenology (see for example, Vogel & Broverman, 1964). Giannitrapani (1985) found reliabilities between .3 and .6 for EEG measures varying in length. O'Gorman and Lloyd (1985) found that a substantial amount (89%) of the variance in EEG measures was associated with person X condition and person X session interactions. In other words, people change differentially (i.e., unreliably) over conditions and over sessions. While these problems are not unresolvable, they make EEG difficult to use easily.

By far the most extensive work on the relationship between IQ and EEG has been done by Giannitrapani (1969, 1985). In this work (Giannitrapani, 1985) children between 11 and 13 were given EEGs and the results were correlated with performance on the Wechsler Intelligence Scale for Children (WISC). A major finding was that 13-Hz activity recorded in the central area was most predictive of intelligence ($r > .48$). Other frequency bands did not show as high a relationship. This finding is corroborated by earlier research done by Gasser, Von Lucadon-Muller, Verleger, & Bucher (1983) in a comparison of mentally retarded persons to a normal control group.

Besides looking at frequency bands, Giannitrapani also correlated WISC subscore patterns with areas of the brain. Pre- and midfrontal and post-temporal areas were most highly associated with differences in WISC patterns.

What do these findings tell us about intelligence? Unfortunately, neither the EEG nor intelligence is well enough understood to answer that question. As with John's technique, it is clear that the electrical activity of the brain is related, in a statistically significant way, to intelligence. The fact that certain frequency bands are most closely associated with intelligence differences suggests that there is an optimal activity level for the brain. Evidently, the most active brain is not the smartest.

C. Average Evoked Potentials

AEPs are much different than EEGs. The recording techniques are similar. Both record electrical activity from the surface of the brain. In AEPs, however, the recording is keyed to a stimulus event. In the first studies of AEPs, a simple visual stimulus was presented to the subject. Electrical activity of the brain was then recorded from one or more electrodes positioned on the surface of the scalp. Many trials were presented using the identical format. Since data are keyed to the stimulus event, all of the trials can be averaged millisecond by millisecond using the stimulus event as the starting point. Averaging eliminates the noise present on each trial and leaves a characteristic waveform. The wave is marked by phases of positive and negative electrical activity. Portions of this wave are named for the time after stimulus presentation that they normally occur and for their valence. For example, "P300" means the characteristically positive portion of the wave that usually occurs 300 ms after the stimulus is presented. Waves recorded under certain standard procedures have such a common, invariant form. Because the form is dependable, a peak that occurs 400 ms after stimulus presentation would be called the P300 peak if the general form of the wave is consistent with the form usually found. In other words, the waveform is thought to represent invariant mental processes that occur even if the time taken for those processes is stretched out.

AEPs are both internally consistent and reliable over time. AEPs have a heritability of .48, about the same as the heritability for intelligence (Osborne, 1970). An AEP is divided into portions. The first 100 ms are called early components. The first 10 ms of the early component represent basic neural mechanisms that are well understood and are used for diagnostic purposes. The remaining portions of the AEP are not well understood either in terms of where they originate or what their function is. All components occurring after 100 ms are called late components. Components after 100 ms are not well understood, either. There has been substantial speculation about the function of the major components (e.g., Donchin & Coles, 1988). The late components have been studied most intensely to find an AEP component linked to intelligence.

First reports of a relationship between AEPs and IQ were by Ertl and his colleagues (Chalke & Ertl, 1965; Ertl, 1971; Ertl & Schafer, 1969). In these studies, P300 latency was correlated with IQ. (Remember that the P300 may not occur exactly 300 ms after stimulus presentation for a subject. Latency of the P300 peak is a measure of when this characteristic peak occurs.) The correlation found ranged from −.3 to −.9. Negative correlations mean that subjects with longer P300 latencies have lower IQs. The problem with Ertl's studies was that subject selection was informal, frequently resulting in extended ranges.

Shucard and Horn (1972, 1973) replicated Ertl's work with more modest results. Most of the correlations between AEP parameters and IQ were around −.15. Shucard and Horn found that correlations got smaller when age effects were removed and larger when the task demanded less attention. Some investigators failed to find any relationship (Rust, 1975). The general conclusion from the many studies that have been conducted in this area is that there is a small to moderate relationship.

Much effort with AEPs has been to refine the measures used. Perry, McCoy, Cunningham, Falgout, and Street (1976) used factor analysis to derive measures of AEPs and related those to IQ using multiple regression. They found moderate correlations. Crawford (1974) used canonical correlations and found smaller correlations. Flinn, Kirsch, and Flinn (1977) used Fourier analysis. The lowest frequencies showed the highest negative correlations between AEPs and IQ.

The derivative measure that has received the most attention is the *string* measure developed by the Hendrickson (Blinkhorn & Hendrickson, 1982; A. E. Hendrickson, 1982; D. E. Hendrickson, 1982; Hendrickson & Hendrickson, 1980). The string measure was initially obtained by lying a string on top of the AEP wave. The length of the string required to trace the wave was the measure recorded. What this measure probably reflects is the variability of the waveform. All waves had the same general shape but the Hendricksons noted that higher IQ persons had more elaborate waveforms. They reasoned that neural transmission of higher IQ persons is more accurate and less subject to distortion or failure than in low-IQ persons. Therefore, the AEP of high-IQ subjects should be more elaborated, reflecting the greater amounts of information being transmitted. In support of this hypothesis, the Hendricksons found a correlation of about .80 between the string measure and IQ.

The findings have been replicated by several investigators (Federico, 1984a, 1984b; Haier, Robinson, Braden, & Williams, 1984; Robinson, Haier, Braden, & Krengel, 1984) who obtained lower correlations. Haier and colleagues found that the effect was largely due to variability in the period between N140 and P200.

A study by Jensen, Schaefer, and Crinella (1981) looked at the relationship between AEPs, IQ, and cognitive tasks. The cognitive tasks were choice reaction time (RT) tasks. Since AEPs measure the speed of central processing it might be expected that an individual's RT would be correlated with their AEP speed. The measure of AEP used in this study was called neural adaptability. Neural adaptability was defined as the difference in AEP waveforms for expected and unexpected events. Neural adaptability was correlated .31 with intelligence. RT tasks are known to correlate reliably with intelligence. When RT measures were combined with neural adaptability to predict IQ, the multiple correlation was .64. Evidently RT and neural adaptability at least partially measure different things. Differences in RT cannot be accounted for by the neural measures used in this study. This study is an important one because it used an approach that must become more common. The study attempted to relate a behavioral measure (choice RT) to an underlying biological measure (neural adaptability) within a theoretical framework. Explaining basic behavioral tasks that relate to intelligence will be important for understanding the biological basis of intelligence.

Following Jensen, Schaefer, and Crinnella's work, a more fine-grained analysis of IQ, AEPs, and RT was carried out by McGarry-Roberts, Stelmack, and Campbell (1992). Using factor analytic measures of P300, they found a correlation of −.36 with general intelligence factor scores. The authors suggest that this correlation results because both P300 and choice RT reflect the stimulus evaluation time and response selection time of cognitive tasks. These general stimulus and response processes vary inversely with intelligence. The kind of information provided by studies that look at theoretically meaningful relationships are likely to be more important than those that find unexplained empirical relationships even if the relationships are large.

AEPs show some promising relationships with IQ. Unfortunately, AEP measures are not well understood. Further, in such complex data there is always the potential for confoundings. For example, it is possible that lower IQ subjects move more during testing creating movement artifacts. Such artifacts might produce a correlation with IQ. Possibilities like this have not been thoroughly explored.

D. Speed of Conduction

An interesting finding by Vernon and Mori (1992) concerns the relationship between peripheral nerve conduction velocity (PNCV) and IQ. PNCV is measured by applying a current at one point on a peripheral nerve (usually in the arm) and measuring how long it takes the current to travel a measured distance down the nerve. PNCV is difficult to measure accurately. Besides requiring precise measurement and calibration, results are affected by minor variations in temperature. When PNCV is measured appropriately the results can show substantial reliable individual differences.

In two studies, Vernon and Mori found correlations of over .40 between PNCV and IQ. PNCV also correlated with measures of RT, but not as highly. Reed and Jensen (1991) did not find the same effects though the techniques used were different. They did find a similar relationship to that found by Vernon and Mori when AEP techniques were used to measure nerve conduction velocity within the brain.

Though a firm conclusion will require further replication of PNCV effects, the results obtained suggest that there are characteristics of nerve conduction that may account for part of intelligence. Considering all of the factors that could affect speed of conduction from myelinezation to transmitter production, it would not be surprising to find that conduction speed was important in intelligence.

VI. IMAGING TECHNOLOGIES

Imaging technologies are, in many ways, the most exciting developments in trying to understand the biological basis of intelligence. They provide a way to look at the brain. The pictures that they produce can show structural characteristics of the brain in living humans. Some techniques can even show the brain at work.

The techniques that already exist will undoubtedly become cheaper and more readily available for research purposes. They will also become more refined and faster. Though techniques now available sounded like science fiction a few years ago, more sophisticated techniques than now available are being developed and will be in use shortly.

A. Magnetic Resonance Imaging

As discussed in the section on gross anatomy, MRI can be an extremely valuable technique for studying anatomical structure in living organisms. MRI produces a picture of any structure showing differences in tissue density. For investigating the brain, pictures taken by MRI can be used to locate anatomical structures, to assess degree of physical damage to the brain, and to assess change over time. MRI is a noninvasive technique. It requires very strong magnetic fields so magnetic metals cannot be brought into the room where the machine is operating. Obviously, persons with magnetic metal plates are not appropriate subjects for the MRI.

MRI can be used to assess blood flow. Since blood flow is related to activity in the region of the brain where it is measured, MRI can be used to assess relative activity level of various parts of the brain. Until recently, other kinds of techniques were more suitable for measuring brain activity because the MRI took a long time to take each picture. New MRI machines can take five pictures or more per second. Temporal resolution of this order may soon make it possible to take what will essentially be moving pictures of the brain while it does cognitive tasks. The pictures that result will be functional pictures that will show which areas of the brain are most active at different stages of the task being done.

B. Positron Emission Tomography

A technique that can show the functional activity of the brain is PET. In this technique, a subject is given a radioactive isotope. During a period set by the half-life of the isotope, the subject does a task given by the experimenter. The length of the uptake period can be from a few minutes to 30 min or more.

All of the isotopes used in PET studies are involved in metabolism in a way that is quantitatively exactly known. Active areas of the brain have a metabolic rate that is higher than when the same area is at rest. Areas of the brain that are most active will absorb the most radioactive isotope. After the uptake period, the subject is placed in a ring of sensors that measure the by-products of the decay of the radioactive isotopes and calculate where the emitted particles originated. The data are accumulated for the entire brain by sections or slices.

What results is a picture of the brain that amounts to a time exposure of metabolic activity during the uptake period. Parts of the brain that were the most active during the uptake task will register the highest levels of radio-active isotope by-products. From a PET study, it is possible to see how active a person's whole brain was and what areas of the brain had the greatest relative activity. Temporal resolution for PET will never be less than the uptake period limited to, at fastest, seconds. This means PET studies will never capture mental tasks while separate processes are being done. Cognitive processes are completed in milliseconds not seconds, which is the temporal resolution of PET. The PET image will produce a record of activity for all processes in the task superimposed on one another as in a photographic time exposure.

Despite the difficulties with PET, they have produced fascinating results. Posner, Petersen, Fox, and Raichle (1988) have used PET to attempt to localize cognitive operations in the human brain. More germane for the relationship between the brain and IQ, Haier et al. (1988) studied PET images of subjects who, during the uptake period, had taken the Raven Progressive Matrices, a test of nonverbal intelligence. The results were counterintuitive.

Haier et al. found that high-IQ subjects had lower overall metabolic rates than subjects with low IQ. This finding suggests that high-IQ subjects didn't have to work as hard to solve the same problems as the low-IQ subjects. Or perhaps high-IQ subjects worked more efficiently. There were no startling findings identifying which areas of the brain were most involved with doing the Raven test.

To follow up on these findings, Haier, Siegel, Tang, Abel, and Buchsbaum (1992) had subjects practice the game, Tetris, over a 1-month period to see if metabolic rates would decline as subjects learned the task. Tetris is a game that appears to require visual-spatial and motor abilities. At the beginning of the experiment each subject took an intelligence test and a PET was taken after an uptake period where subjects played Tetris. Subjects then practiced Tetris for 30–45 min for about 1 month (average of about 19 practice sessions for each subject). At the end of the practice period, a PET was taken after each subject had played Tetris during the uptake period.

Because PETs had been taken both before and after practice, comparing the two scans showed the brain areas where metabolic activity changed most with practice. Subjects with the highest IQ showed the largest decline in metabolic activity in areas that changed with practice. Subjects with the highest IQ also showed the largest decline in total metabolic rate. Both findings support the conclusion from the original study. Subjects with the highest IQ show the largest decline in activity as they become practiced. Haier has suggested an efficiency hypothesis to explain the results from these PET studies: People with the highest IQ have the most efficient brains.

PET studies have just begun to be done. They offer the opportunity to watch the human brain in action. Behavioral scientists have never had the opportunity to see a brain behaving. PET studies are not perfect and have many methodological problems associated with them. They provide information about the relationship between the brain and intelligence that could only be guessed about before.

VII. GENETICS

There is substantial and undeniable indirect evidence of the effects of genetic factors on intelligence (Plomin, De Fries, & McClearn, 1990). Most of this evidence comes from studies of persons who differ in degree of relationship and who have been reared in different ways. Adoption studies, twin studies, and studies of twins reared apart, among other kinds of studies, contribute to the conclusion that the heritability of intelligence is between .4 and .8. Based on this evidence, there can be little doubt that there is a hereditary basis for intelligence. All the evidence leading to this conclusion is indirect, though.

Current techniques can provide more direct evidence. It is now possible to study allele frequency at particular sites on a gene (McClearn, Plomin, Gora-Maslak, & Crabbe, 1991). When persons differing in intelligence differ in the allele frequency at a particular location on a fragment of a chromosome, it is evidence that the gene location or one close to it on the fragment is implicated in intelligence. The particular gene that contributes to intelligence can then be located. The function of the particular alleles implicated can be studied to see exactly how they influence intelligence. Studies of this kind are currently under way. This kind of research has already provided valuable information about alcoholism in humans. Alcoholism is more difficult to measure reliably in humans than is intelligence. It should be expected that intelligence, because of its reliable measurement and substantial heritability, will yield more easily to this kind of analysis than most other polygenic human traits.

The problem with studies looking for genes implicated in polygenic traits like intelligence is that there are many genes. Maybe half of them are implicated in brain functioning. Finding the location of particular genes that have large effects on intelligence will be a little like looking for a needle in the proverbial haystack. Of all human polygenic traits, intelligence can be most reliably measured and shows the largest genetic contribution. These desirable characteristics will make it easier to locate specific genes affecting intelligence than for other polygenic behaviors, but it still may be a difficult search.

Being able to trace the path from a single gene to its effect on complex human behavior will undoubtedly change the way we think of intelligence. If research is successful in identifying exactly what genes contribute to differences in intelligence, for the first time we will have an exact idea about the biological origins of a behavioral trait. Such knowledge will change the way we think about the biological basis of behavior. Fifty years from now we may find current speculations about the biological basis of intelligence to be naive and entirely wrong.

VIII. CONCLUSIONS

The examination of the relationship between intelligence and the brain raises many questions. These questions have been asked for as long as intelligence has been measured. There may never be an ultimate answer to any of these questions. The answers will have to be modified as we slowly expose the truth about intelligence. In the following sections, I attempt to provide the best answer available to some of these questions based on current evidence.

A. Is There a Brain–IQ Relationship?

In one sense, this is a trivial question. All psychological processes must have a biological basis in the nervous system. The brain is undoubtedly important in carrying out higher mental processes. There is also no doubt that all measurable differences in intelligence should be present in biological processes. That is, if we had complete knowledge about a person's biological brain processes, we could perfectly predict the person's score on an IQ test. All this statement says is that both biological and environmental effects on intelligence must be instantiated in biological processes. There is no other possibility short of postulating some form of mental parallelism or mind independent of the brain.

Is there a relationship between the brain and intelligence? The simple answer is yes. The empirical evidence for the conclusion is overwhelming. If there is no distinction between whether the biological differences arise from biological limitations or environment, there can be no argument that the brain and intelligence are intimately related.

B. Is There Empirical Evidence of a Brain–IQ Relationship?

What most people are really concerned about when they discuss the biological basis of intelligence is whether each person's unique biological characteristics place a limit on intelligence. The evidence for biological limitations is currently most indirect. Some of it has not been discussed here. Mental retardation often results from impaired biological processes that limit intellectual development. Evidence from behavioral genetic studies shows that a substantial portion of intelligence is heritable. Perhaps the most convincing evidence of all is that vigorous attempts to raise IQ have only had limited effectiveness (Spitz, 1986). All of these sources of evidence are indirect.

With new techniques to study biological processes in living humans, there will be increasing opportunity to find direct evidence of the biological basis for individual differences in intelligence. Previous sections of this chapter presented studies that signal what will almost certainly be an explosion of studies addressed at this issue. The studies presented here are really first examples of what will follow.

Even these preliminary studies provided some important evidence about the biology that may produce individual differences. Sex differences is one area where a convincing argument is being made for biological differences affecting intellectual processes. Other areas of inquiry also provide preliminary data that suggest differences in intelligence could be related to basic biological differences.

Ultimate answers come down to what is acceptable as definitive evidence. If an acceptable answer requires tracing the linkage from an individual gene to differences in behavior, then the question is still unanswered. If, on the other hand, more indirect evidence is acceptable, then there is little question of the link between individual differences in intelligence and biological differences.

C. Where Are the Largest Research Gains Likely to Come?

Predicting which of the new techniques for studying the relationship between biology and intelligence will produce the biggest discoveries is risky because of the rate of technological change. Many newer techniques owe their existence to the computer chip. Imaging and electrophysiological techniques require the manipulation of huge amounts of data that would be impossible without fast computers. As computers become better and faster, the techniques will improve. How good and how cheap they will get is hard to know. It is fully conceivable that these kinds of computer-driven techniques will revolutionize the study of human intelligence.

Another area where inevitable progress awaits is the study of the direct genetic contribution to intelligence. Finding even one or a few genes that contribute to differences in human intelligence, even if that contribution is small, will have important consequences for theories of intelligence. First, it will end the debate about the possibility of a genetic contribution to human intelligence. Second and more important, it will give an idea of what biological factors can affect intelligence. Will it be genes that regulate receptors? Will it be genes that program the nervous system's development? Just what kinds of genes affect intelligence? How many of them are there? How much variance does each gene account for?

In the end, it will not be a single method of study or a single technique that tells us all we need to know about differences in human intelligence. It will be the combined and coordinated information from all available techniques that brings understanding. Because of the explosion of new methods to study the biology of intelligence, these are bound to be very exciting and productive times.

D. Of What Importance Are the Behavioral Sciences in This Work?

Though many techniques used to study the biological basis of human intelligence have been developed in biology and medicine, their application will depend importantly on the behavioral sciences. To relate behavior to brain functioning, the behavior has to be well defined and accurately measured. If the behavior cannot be defined reliably, relating it to biological structures will be impossible. For that reason, intelligence is a natural behavior to begin to relate to individual differences in biology. It is reliable and has been extensively studied. Those who study intelligence will make natural collaborators with those who have developed methods for measuring brain functioning.

ACKNOWLEDGMENTS

Parts of this work were supported by Grants No. HD07176 and HD15516 from the National Institute of Child Health and Human Development, Office of Mental Retardation, the Air Force Office of Scientific Research and the Brooks Air Force Base Human Resources Laboratory, Project Lamp.

REFERENCES

Ankney, C. D. (1992). Sex differences in relative brain size: The mismeasure of woman, too. *Intelligence, 16,* 329–336.

Beach, F. A., Hebb, D. O., Morgan, C. T., & Nissen, H. W. (1960). *The neuropsychology of Lashley.* New York: McGraw-Hill.

Blinkhorn, S. F., & Hendrickson, D. E. (1982). Averaged evoked responses and psychometric intelligence. *Nature, 295,* 596–597.

Chalke, F. C. R., & Ertl, J. P. (1965). Evoked potentials and intelligence. *Life Sciences, 4,* 1319–1322.

Crawford, C. B. (1974). A canonical correlation analysis of cortical evoked response and intelligence test data. *Canadian Journal of Psychology, 28,* 319–332.

Detterman, D. K. (1987). Theoretical notions of intelligence and mental retardation. *American Journal of Mental Deficiency, 92,* 2–11.

Donchin, E., & Coles, M. G. H. (1988). Precommentary: Is the P300 component a manifestation of context updating? *Behavioral and Brain Sciences, 11,* 355–372.

Ertl, J. P. (1971). IQ, evoked responses, and Fourier analysis. *Nature, 241,* 209–210.

Ertl, J. P., & Schafer, E. W. P. (1969). Brain response correlates of psychometric intelligence. *Nature, 223,* 421–422.

Eysenck, H. J. (1986). Toward a new model of intelligence. *Personality and Individual Differences, 7,* 731–736.

Federico, P. A. (1984a). Event-related-potential (ERP) correlates of cognitive styles, abilities and aptitudes. *Personality and Individual Differences, 5,* 575–585.

Federico, P. A. (1984b). Hemispheric asymmetries: Individual-difference measures for aptitude-treatment interactions. *Personality and Individual Differences, 5,* 711–724.

Fischbach, G. D. (1992). Mind and brain. *Scientific American, 267*(3), 48–57.

Flinn, J. M., Kirsch, A. D., & Flinn, E. A. (1977). Correlations between intelligence and the frequency content of the visual evoked potential. *Physiological Psychology, 5,* 11–15.

Gasser, T., Von Lucadon-Muller, I., Verleger, R., & Bucher, P. (1983). Correlating EEG and IQ: A new look at an old problem using computerized EEG parameters. *Electroencephalography and Clinical Neurophysiology, 55,* 493–504.

Giannitrapani, D. (1969). EEG average frequency and intelligence. *Electroencephalography and Clinical Neurophysiology, 27,* 480–486.

Giannitrapani, D. (1985). *The electrophysiology of intellectual functions.* New York: Karger.

Grafman, J., Jonas, B. S., Martin, A. Salazar, A., Weingartner, H., Ludlow, C., Smutok, M. A., & Vance, S. C. (1988). Intellectual function following penetrating head injury in Vietnam veterans. *Brain, 111,* 169–184.

Grafman, J., Salazar, A., Weingartner, H., Vance, S., & Amin, D. (1986). The relationship of brain-tissue loss volume and lesion localization to cognitive deficit. *The Journal of Neuroscience, 6,* 301–307.

Green, E. J., & Greenough, W. T. (1986). Altered synaptic transmission in dentate gyrus of rats reared in complex environments: Evidence from hippocampal slices maintained in vitro. *Journal of Neurophysiology, 55,* 739–749.

Green, E. J., Greenough, W. T., & Schlumpf, B. E. (1983). Effects of complex or isolated environments on cortical dendrites of middle-aged rats. *Brain Research, 264,* 233–240.

Greenough, W. T., McDonald, J. W., Parnisari, R. M., & Camel, J. E. (1986). Environmental conditions modulate degeneration and new dendrite growth in cerebellum of senescent rats. *Brain Research, 380,* 136–140.

Haier, R. J., Robinson, D. L., Braden, W., & Williams, D. (1984). Evoked potential augmenting-reducing and personality differences. *Personality and Individual Differences, 5,* 293–301.

Haier, R. J., Siegel, B. V., Nuechterlein, K. H., Hazlett, E., Wu, J. C., Paek, J., Browning, H. L., & Buchsbaum, M. S. (1988). Cortical glucose metabolic rate correlates of abstract reasoning and attention studied with positron emission tomography. *Intelligence, 12,* 199–217.

Haier, R. J., Siegel, B., Tang, C., Abel, L., Buchsbaum, M. S. (1992). Intelligence and changes in regional cerebral glucose metabolic rates following learning. *Intelligence, 16,* 415–426.

Hebb, D. O. (1949). *The organization of behavior: A neuropsychological theory.* New York: John Wiley & Sons.

Hebb, D. O. (1980). The structure of thought. In P. W. Jusczyk & R. M. Klein (Eds.), *The nature of thought: Essays in honor of D. O. Hebb* (pp. 19–36). Hillsdale, NJ: Erlbaum.

Hendrickson, A. E. (1982). The biological basis of intelligence. Part I: Theory. In H. J. Eysenck (Ed.), *A model of intelligence* (pp. 151–196). New York: Springer-Verlag.

Hendrickson, D. E. (1982). The biological basis of intelligence. Part II: Measurement. In H. J. Eysenck (Ed.), *A model for intelligence* (pp. 197–228). New York: Springer-Verlag.

Hendrickson, D. E., & Hendrickson, A. E. (1980). The biological basis of individual differences in intelligence. *Personality and Individual Differences, 1,* 3–33.

Hubel, D. H., & Wiesel, T. N. (1970). The period of susceptibility in the physiological effects of unilateral eye closure in kittens. *Journal of Physiology, 206,* 419–436.

Huttenlocher, P. R. (1974). Dendritic development in neocortex of children with mental defect and infantile spasms. *Neurology, 24,* 203–210.

Huttenlocher, P. R. (1979). Synaptic density in human frontal cortex-developmental changes and effects of aging. *Brain Research, 163,* 195–205.

Huttenlocher, P. R. (1984). Synapse elimination and plasticity in developing human cerebral cortex. *American Journal of Mental Deficiency, 88,* 488–496.

Huttenlocher, P. R., deCourten, C., Garey, L., & Van der Loos, H. (1982). Synaptogenesis in human visual cortex: Evidence for synapse elimination during normal development. *Neuroscience Letters, 33,* 247–252.

Jensen, A. R., Schaefer, E. W., & Crinella, F. M. (1981). Reaction time, evoked brain potentials, and psychometric 'g' in the severely retarded. *Intelligence, 5,* 179–197.

John, E. R. (1977). *Neurometrics: Clinical application of quantitative electrophysiology.* Hillside, NJ: Erlbaum.

John, E. R., Karmel, B. Z., Corning, W. C., Easton, P., Brown, D., Ahn, H., John, M., Harmony, T., Prichep, L., Toro, A., Gerson, I., Bartlett, F., Thatcher, R., Kaye, H., Valdes, P., & Schwartz, E. (1977). Neurometrics. *Science, 196,* 1393–1409.

Lashley, K. S. (1929). *Brain mechanisms and intelligence.* Chicago, IL: University of Chicago Press.

McClearn, G. E., Plomin, R., Gora-Maslak, G., & Crabbe, J. C. (1991). The gene chase in behavioral science. *Psychological Science, 2,* 222–229.

McGarry-Roberts, P. A., Stelmack, R. M., Campbell, K. B. (1992). Intelligence, reaction times, and event related potentials. *Intelligence, 16,* 289–313.

O'Gorman, J. G., & Lloyd, J. E. M. (1985). Is EEG a consistent measure of individual differences? *Personality and Individual Differences, 6,* 273–275.

Osborne, R. T. (1970). Heritability estimates for the visual evoked response. *Life Sciences, 9*(Part II), 481–490.

Perry, N. W., McCoy, J. G., Cunningham, W. R., Falgout, J. C., & Street, W. J. (1976). Multivariate visual evoked response correlates on intelligence. *Psychophysiology, 13,* 323–329.

Plomin, R. R., DeFries, J. C., & McClearn, G. E. (1990). *Behavioral genetics: A primer* (2nd edition). New York: W. H. Freeman and Company.

Posner, M. I., Petersen, S. E., Fox, P. T., & Raichle, M. E. (1988). Localization of cognitive operations in the human brain. *Science, 240,* 1627–1631.

Purpura, D. P. (1975). Dendritic differentiation in human cerebral cortex: Normal and aberrant developmental patterns. *Advances in Neurology, 12,* 91–116.

Reed, T. E., & Jensen, A. R. (1991). Arm nerve conduction velocity (NCV), brain NCV, reaction time, and intelligence. *Intelligence, 15,* 33–47.

Robinson, D. L., Haier, R. J., Braden, W., & Krengel, M. (1984). Psychometric intelligence and visual evoked potentials: A replication. *Personality and Individual Differences, 5,* 487–489.

Rosenzweig, M. R., & Bennett, E. L. (1977). Effects of environmental enrichment or impoverishment on learning and on brain values in rodents. In A. Oliverio (Ed.), *Genetics, environment, and intelligence.* Amsterdam, Netherlands: Elsevier.

Rushton, J. P. (1992). Cranial capacity related to sex, rank, and race in a stratified random sample of 6,325 U.S. military personnel. *Intelligence, 16,* 410–413.

Rust, J. (1975). Cortical evoked potential, personality, and intelligence. *Journal of Comparative and Physiological Psychology, 89,* 1220–1226.

Schucard, D. W., & Horn, J. L. (1972). Evoked potential amplitude change related to intelligence and arousal. *Psychophysiology, 10,* 445–452.

Shucard, D. W., & Horn, J. L. (1973). Evoked cortical potentials and measurement of human abilities. *Journal of Comparative and Physiological Psychology, 78,* 59–68.

Spitz, H. H. (1986). *The raising of intelligence: A selected history of attempts to raise retarded intelligence.* Hillsdale, NJ: Lawrence Erlbaum Associates.

Stott, D. H. (1983). Brain size and "intelligence." *British Journal of Developmental Psychology, 1,* 279–287.

Thompson, R., Crinella, F. M., & Yu, J. (1990). *Brain mechanisms in problem solving and intelligence: A lesion survey of the rat brain.* New York: Plenum Press.

Vernon, P. A., & Mori, M. (1992). Intelligence, reaction times, and peripheral nerve conduction velocity. *Intelligence, 16,* 273–288.

Vogel, W., & Broverman, D. M. (1964). Relationship between EEG and Test Intelligence: A Critical Review. *Psychological Bulletin, 62,* 132–144.

Weinstein, S., & Teuber, H. L. (1957). Effects of penetrating brain injury on intelligence test scores. *Science, 125,* 1036–1037.

Willerman, L., Schultz, R., Rutledge, J. N., & Bigler, E. D. (1992). Hemisphere size asymmetry predicts relative verbal and nonverbal intelligence differently in the sexes: An MRI study of structure–function relations. *Intelligence, 16,* 315–328.

Wood, C. C. (1978). Variations on a theme by Lashley: Lesion experiments on the neural model of Anderson, Silverstein, Ritz, & Jones. *Psychological Review, 85,* 582–591.

Wood, C. C. (1980). Interpretation of real and simulated lesion experiments. *Psychological Review, 87,* 474–476.

Yeo, R. A., Turkheimer, E., Raz, N., and Bigler, E. D. (1987). Volumetric asymmetries of the human brain: Intellectual correlates. *Brain and Cognition, 6,* 15–23.

4

The Neuropsychology of Sex-Related Differences in Brain and Specific Abilities

Hormones, Developmental Dynamics, and New Paradigm

Helmuth Nyborg

I. INTRODUCTION

The neuropsychological study of sex differences in specific mental abilities is a minefield of methodological and theoretical problems. We are faced with the paradox that neuropsychology abounds with empirical data and detailed clinical observations but lacks a unifying theory, whereas the study of sex differences abounds with (typically mutually exclusive) theories and correlation coefficients but is meager on good hard data. Smooth integration of neuropsychology with sex differences research is further made difficult by fundamental disagreement about possible reason(s) for differences in specific abilities, and because various putative reasons have not yet materialized into precise operational definitions that allow for formal testing in tight experimental designs. The final and perhaps most critical hurdle for progress is that sex researchers often lean exclusively on group mean data, whereas clinical neuropsychologists typically harvest idiographic information in case studies. Another line of research, psychometrics, often collects data on sex differences in specific abilities and typically uses factor analytic or partial correlation coefficient techniques. These approaches have yet to prove their value as a platform for testing causal hypotheses about brain mechanisms behind specific abilities. The net result of all this is that we have laboriously acquired a huge amount of questionable data about average sex differences in specific abilities and are in urgent need of accurate theoretical interpretation and meaningful cross-disciplinary integration.

The main thesis of this chapter is that we now need a completely new, coherent, and integrative paradigm in the neuropsychological study of individual and sex differences in specific abilities. The tremendous recent advances in understanding the nuts and bolts of developmental neuroendocrinology suggest that this area holds potential for accomplishing such an ambitious goal. The past 10–20 yr of research in this field has thus produced new and often quite surprising evidence, which promises proper identification of the proximate base(s) for sexual differentiation. It seems possible within this field to tentatively define mechanisms and biological loci of action, and to apply this knowledge outside the field in order to derive rather precise and testable hypotheses about person-specific brain-ability relationships. Obviously, such a radical cross-disciplinary approach presupposes a readiness to accept that sex differences in specific abilities can be handled in ways that differ fundamentally in some respects from the usual psychological, genetic, and psychometric treatment. It presupposes, in other words, methodological as well as theoretical reorientation.

The chapter is divided into two parts. The first part is a selective review of observations made in accordance with the traditional sex-dimorphic approaches, and focuses on problems with these approaches. It is concluded that there probably will be little progress before adequate solutions are found to the following number of problems. It seems to have been a dubious procedure from the beginning to first calculate separate averages for male and female abilities, and then to test for statistical differences between the averages. There is little agreement about how to identify and operationalize the causal agent(s) behind sexual differentiation. The predictive power of specific ability scores for performance outside the test room is modest at best. More importantly, traditional theories of sexual differentiation cannot account for recent evidence of very dynamic gene–experience–brain–ability relationships. The aim of the second part of the chapter is to try and implement a paradigmatic shift in the neuropsychological study of sex differences in specific abilities. A physicochemical model for body and brain development is presented, and its theoretical and evolutionary framework is discussed. It is argued, that the model and the theory ordain revision of basic assumptions behind the traditional nature–nurture interaction model.

II. PART I: A SELECTIVE REVIEW OF SEX-DIMORPHIC STUDIES OF SPECIFIC MENTAL ABILITIES

One of the first written accounts of a sex difference in specific mental abilities is found in an ancient Sanskrit book, in which we are told that "Ten

shares of talk were handed down to Earth—the nine went to the women."
This early view is repeated in more recent and systematic reviews of quantitative sex differences in mental abilities. Garai and Scheinfield (1968) and
Maccoby and Jacklin (1978) thus suggest that females surpass males in some,
but not necessarily in all, areas of verbal ability. With respect to development,
girls speak on average earlier and better than boys, and boys are over-rep-
resented among stutterers, dyslexics, and general learning-disabled children.
Females further outperform males on delicate fine-motor tasks and in some
tasks requiring speeded perceptual scanning, whereas males outperform females in gross-motor tasks like throwing accuracy, even when no great force
is called for (Jardine & Martin, 1983; Watson & Kimura, 1990), a difference
that is better considered in an evolutionary perspective (Kolakowski &
Malina, 1974).

One of the most pronounced and robust sex differences is seen in spatial
abilities (McGee, 1979), and in the many activities assumed to draw upon
good spatial skills. Unfortunately, despite wholehearted efforts (e.g., Smith,
1964) we still have no clear-cut definitions of the various spatial abilities and
their interrelations. Nevertheless, however defined, males on average outperform females in most standard and also in not-so-standard tests for spatial
abilities. Males excel in simple recollection, visual coding, and disembedding
of geometrical shapes and figures, in mental rotation and identification tasks,
in geometrical and in particular in mathematical problem-solving tests (but
not in numerical tasks), in chess, in tests of direction sense, in visual and
walking maze tests, in tactual mazes, in pattern walking, in map reading, in
absolute direction tasks, in left–right discrimination, in aiming and tracking,
in the rod-and-frame test and embedded-figures test, in geographical knowledge, in Piagetian three-dimensional tasks, in logical conservation, in the
water-level test, and in music composition. Harris (1979) has provided an
instructive overview of these sex differences. Males also dominate in educational and occupational areas believed to involve spatial abilities, such as
architecture, physics, and engineering. It is worth remembering, however,
that specific ability scores usually correlate only moderately with achievement
in these areas. Finally, at the extreme ends of ability dimensions, there are
more geniuses among males than females, but also more idiots. Male abilities
tend, in other words, to vary more than female abilities.

Many more differences could be lined up here, but this is not necessary for
the present purpose. First, excellent and fairly exhaustive reviews of sex
differences are published regularly. More importantly, however, the sex differences already mentioned serve well to illustrate two major points, namely,
that they mirror fundamental methodological and theoretical problems in the
field, and that these difficulties prevent real progress in the neuropsychological study of sex differences in abilities. I shall first discuss the methodological
problems, and then address some serious theoretical problems.

A. Methodological Problems

1. Introduction At least five major methodological problems impair sex-dimorphic studies of specific abilities. The first is that genetic, genital, or self-reported sex is a dubious classificatory variable for abilities. Exclusive reliance on male–female averages tends to obscure the considerable individual variation, and leaves out important information about individual differences. Second, it is a truism to say that the choice of research methods depends on the nature of the phenomenon to be studied. However, with respect to sex differences there is widespread confusion about the cause. This, in turn, makes it difficult to decide on the best (and probably only—see later) method for examining details of the cause(s) of the differences. Third, most specific ability tests show only moderate sex differences, but even a quick glance by the untrained eye indicates that men and women achieve quite differently in all known societies in areas believed to draw heavily on specific abilities, whether in education, occupation, or in top positions in the corporate or political power structures. Can we take the low correlation to mean that specific ability measures are inherently unreliable, or that they are so narrow in scope that they must be supplemented by further information from other areas related to specific abilities to be usable? Fourth, sex differences in specific abilities vary with age. This means that studies failing to consider neuropsychological sex differences in specific abilities within a developmental framework may be at fault. Fifth, ideology and concern over "political correctness" have a significant impact on research on sex differences. This is particularly obvious in the United States, where popular sexual policy at times hampers the rights of researchers to work in the area or makes publication difficult. But even in Denmark, considered by many as a politically and sexually most liberal country, it is almost impossible to fund studies of the biological basis of sex differences in abilities, whereas social learning approaches, questionnaires research, and correlation studies are generously rewarded with million-dollar funding in the absence of previous meaningful results. Even the construction of tests for specific abilities has been perverted so as to better conform to prevailing norms for "political correctness" or to the ill-defined notion of sexual equality. I will first discuss each of these problems in some detail, and then suggest methodological and theoretical corrections in order to promote a more adequate scientific program for the neuropsychological study of specific abilities.

2. Sex Dimorphism People seldom find it difficult to identify themselves or others as either male or female. Self-reported, genital, or genetic sex is, accordingly, an easy way of classifying humankind into two groups. There are, however, several very good reasons for not using this procedure when looking for the basis for difference in abilities. It is, for example, a risky business to go the other way around and tell an individual's sex from knowledge about his or her specific ability score, as the distributions of male and female ability scores overlap to a considerable degree for all but a few areas. One possible exception is high-level problem-solving mathematics. Here male superiority increases rapidly with increasing item difficulty, so that there is virtually no overlap at the highest levels of ability (Benbow, 1988). But for most other abilities, the large within-group variability in abilities and the overlap in distributions both decrease the probability of finding statistically significant differences, so that the average sex differences in abilities typically hover around a modest 5% figure. This observation gives some researchers reason to conclude that the scientific or practical value of studying sex differences in abilities is low (e.g., Jacklin, 1979; Plomin & Daniels, 1987). Besides the custom of treating individual variation as statistical "noise," there are other reasons for not accepting this conclusion at face value. Whereas extreme within-sex scores counterbalance each other and moderate the group mean, the whole procedure makes it almost impossible to identify the cause(s) for the differences among individuals (Nyborg, 1977, 1983). To give an example, social learning theorists typically explain female superiority in verbal fluency by general or specific sex differences in upbringing. This intuitively understandable assumption is then tested by correlating average type of upbringing with average level of fluency. It is well known, however, that correlation does not prove causation, but the point here is that the relatively lower male and higher female verbal fluency averages usually cover a few high-ability males and low-ability females. The performance of these individuals is nevertheless "explained" by arguments used to explain also the opposite trend. The sex-dimorphic approach becomes a defective tool for unraveling the underlying individual cause(s) of development of a sex difference in verbal fluency, because it by and large misses exact information about within-sex high- and low-ability differences, and because it is unable to deal with causes except in the most general and abstract sense. The explanation, that girls excel in verbal fluency because they are reared in a particular way is compromised by the fact that some similarly reared girls score low on verbal fluency, and that some presumably differently reared boys score high. Sex dimorphism confounds, in other words, different individuals and masks causes. It might be essentially correct to declare, on the basis of available data, that sex differences in specific abilities are trivial, but the nature of the sex-dimorphic approach makes this statement more or less empty. In the ideology section I shall demonstrate that the statement is also perfectly circular. The relevant question is really whether there are proper ways to identify the sources of the tremendous individual and sex-related differences in abilities. In this manner we better substitute the nomothetic sex-dichotomous approach with a person-specific approach. Only by concentrating on individuals is it possible to avoid inadvertently confounding different people and to enhance the probability of identifying the proper causes for differences in specific abilities.

3. Causal Factors   Obviously, research methods must fit the nature of the task at hand. If our hypothesis is that genes cause differences in abilities, we use specialized methods from molecular, population, or behavioral genetics. But most sex difference researchers employ one or another variation of the social learning approach. They typically inquire into people's experiences by asking them to fill out lengthy questionnaires, or observe them in various social or field situations. They then interpret the observations in terms of the hypothesis, that differences in social experiences, norms, or stereotypes explain differences in abilities. Investigators with a psychometric inclination collect data by test batteries, and then factor analyze them and derive principal and secondary components, or calculate partial correlations. From this they sometimes make inferences about underlying brain processes. Neuropsychologists typically study clinical cases and judge the results on an individual basis or with reference to more or less representative age, sex, or population norms. Aside from the clinical use of data, neuropsychologists also reason about the implications of reduced performance, as a function of circumscribed lesions in particular brain areas, of medication, or of sex-specific anatomical or functional differences in brain function (e.g., hemispheric lateralization), in order to explain sex differences or to develop a theory of brain–ability relationships in general.

Each of these approaches has greatly enlarged our descriptive knowledge, but with respect to a deeper understanding of the reason(s) for the sex differences in abilities none of them can be said to have enjoyed great success. The situation is rather that various approaches tend to produce mutually exclusive results, while not providing a clue as to which explanation is the scientifically more acceptable. We are faced with the ugly dilemma that literally thousands of studies have neither allowed us to operationalize the proper cause(s) of sex differences in specific abilities, nor to decide on the proper methods, interpretation, and integration of the mountain of data. One solution to this is to continue to shoot in all directions, to try and work out better theories for interaction, or to remain elective and combine the best from all approaches. So far none of these strategies promises rapid progress. Perhaps this view may appear unduly pessimistic to many experts in the field. I think, however, that most will agree with me that real progress depends on our ability to identify and operationalize the proximate cause(s) for sexual differentiation of specific abilities. Such a feat would greatly help us in choosing proper methods for further studies, preferably ones that allow for a rigorous falsification procedure.

4. The Developmental Perspective   Everybody with experience in the field of abilities knows how essential it is to keep track of the developmental status of each single individual when studying sex differences. Most sex differences in abilities are rather small before puberty, but multiply within a short time window of a couple of years, take direction, and increase in size, to reach a maximum during the later phases of puberty. Another interesting feature is that the appearance of differences in specific abilities is to some extent harmonized with the maturational tempo of the body (Petersen, 1976; Petersen & Crockett, 1985; Petersen & Taylor, 1980). Several studies have found that early and late maturation relates to different patterns of ability. Early-maturing individuals tend to end up with higher verbal than spatial abilities, and late-maturing individuals tend to display the opposite pattern of development (Crockett & Petersen, 1985; Waber, 1976, 1977a, 1977b). A controversial aspect of the development of abilities is that more females than males seem to actually regress in spatial abilities around puberty (Nyborg, 1988a; Nyborg & Nielsen, 1977; see review by Nyborg, 1983). The female regression has been explained in terms of increasing social pressure towards conformity in puberty, or by accumulation of sex-specific learning. Other theorists thought that the female regression was due to a subject sampling bias (Witkin, Goodenough & Karp, 1967). Few researchers have dared to ask if a biological explanation was appropriate. The model presented later in the chapter allows us to examine this possibility by generating testable predictions about whether particular females run a higher risk of regressing in spatial abilities around puberty. The model also makes predictions about which males are likely to suffer a similar regression.

5. Modest Ability–Achievement Correspondence   Most educations and occupations require different involvement of specific abilities and show differences in the proportion of male and female participation. Some of these differences are large enough to talk about "male" and "female" educations and occupations (Danmarks Statistik, 1985; Ragins & Sundstrom, 1989). Moreover, the strongly biased distribution of males in top positions in almost all areas (Reid, 1982), and in the societal power structure in general (Goldberg, 1977; Wormald, 1982), is too obvious to be easily missed or pronounced trivial. On this background, it is indeed puzzling to realize, that most standard tests show a modest sex difference in the abilities required for achievement. It is equally surprising to learn that metatheoretical analyses of literally hundreds of studies of sex differences in abilities often come out with almost negligible differences (e.g., Linn & Petersen, 1985). As mentioned, these results are taken to mean that sex differences in abilities are "trivially small" (Jacklin, 1979), that sex differences diminish over time (Hyde & Linn, 1986; Jacklin, 1989), or that sex explains less than 5% of the total variability (Plomin & Daniels, 1987). The last mentioned study concluded that the major platform for explaining individual differences in the development of abilities and personality is social and extrafamilial. There are several alternatives to this explanation, however. Perhaps single-ability test scores are less than optimal predictors for later development and achievement. Even though ability tests predict later educational achievement better than teacher evaluation, tests differ with respect to how well they reflect a sex difference in a particular ability dimension, and the ability measured in a particular study may not be entirely relevant for predicting later achievement in related areas. Ability tests may leave out vital information about *how* a given individual makes use of his or her specific abilities outside the test room. The position taken in this chapter is that ability test scores must be supplemented with covariant measures of other characteristics of the individual in question to gain credibility as a predictor variable. Such an idea is neither new nor without problems. The bold attempt by Witkin et al. (1954) and by others from the group around Witkin (Witkin, Dyk, Faterson, Goodenough, & Karp, 1962; Witkin & Goodenough, 1977; Witkin, Goodenough, & Oltman, 1977) to connect spatial ability, field dependence, and cognitive style with personality, within the concept of psychological differentiation, met with only moderate success, even though field dependence measures correlate with particular types of educational choice (Witkin, 1973; Witkin, Moore, Goodenough, & Cox, 1977). A particularly nagging problem has been that personality measures do not usually correlate well with specific ability scores. There may be a solution to these problems, as suggested later in this chapter. The basic problem may be how to first establish causal connections between specific abilities and particular personality traits, and then to relate these to the development and achievement of a single individual. The model to be presented later strives to formalize these connections in terms of a molecular bridge between genes, experience, the development of ability, body, brain, personality, and achievement.

6. Ideology Run Wild   Research on sex differences in specific abilities is a sensitive area, and the ramification of ideology should not easily be dismissed. The field is looked upon by many with suspicion, and not a few believe that the topic is too controversial to be worth the trouble to investigate. Public debates often confuse facts with ideology and "what is" with "what should be," and this sometimes reflects back on science. Personal experience teaches me that public discussions often take the following general forms. There are no sex differences in abilities! OK, there are sex differences, but they are trivial. Well, perhaps some of the differences are important, but they arise from unfortunate socialization practices, are malleable, and will go away with time or a little help from friends. Well, sex differences may be important and genes may be responsible. This only means that we must make an extra effort to change peoples' attitude or society to counter the effects. By the way, do you know how biased ability tests are, and that—what was his name—faked the heritability estimates. In case none of this works, discussants may set unrealistically high scientific criteria for the collection and interpretation of data on sex differences, question the motives or moral integrity of the sex researchers, politicize the observations, or ask not to be bullied, patronized, or discriminated against. In this respect, research on sex and race differences suffers equal fate. At the same time, opponents often present correlations, anecdotal evidence, or strong beliefs as counterevidence. This unsound semi-professional climate for examining sex differences in specific abilities has soured the soil for many studies, such as those by Benbow (1988), Hampson (1986), and Hampson and Kimura (1988). The major reason for mentioning such problems here at all is, however, that the hostile public atmosphere bounces back with measurable negative effects at all levels of sex research. Test constructors have, for example, felt forced to calibrate their apparatus to conform to dogmas of sexual equality. Preliminary versions of Wechsler's well-known IQ tests showed "unacceptable" male and ethnic advantages in overall IQ, but then certain subscale items, believed to be unfair to "certain minority groups" were removed, so that final versions of the test no longer showed a sex difference in overall IQ (Kaufman, 1975; Vogel, 1990; Wechsler, 1981). Conversely, the common female verbal superiority in the Scholastic Aptitude Test (SAT) was moderated by simply adding to the verbal content side areas outside the humanities—such that typically interest males more—and this favored male performance. However, no attempt was made to counter the male advantage in mathematics. These examples illustrate an important point. The "empirical fact" of no sex difference in g is entirely arbitrary. It is based on the ideologically inspired a priori decision that there must be no overall sex difference. It is therefore rather puzzling to see that recent analyses of the American standardization sample of the Wechsler Intelligence Scale for Children—Revised (WISC-R) by Jensen and Reynolds (1983) and of the Scottish standardization sample of the WISC-R by Lynn and Mulhern (1991) both indicate that boys no longer excel in performance IQ (which was expected from previous research), but now obtain a significantly higher full-scale IQ and, surprisingly, also a higher verbal IQ. Matarazzo, Bornstein, McDermott and Noonan (1986) reanalyzed the normative Wechsler Adult Intelligence Scale—Revised (WAIS-R) sample (Wechsler, 1981) and found that males significantly outperformed females on verbal IQ, on performance IQ, and accordingly on full-scale IQ. They find, however, that the differences are trivial, as their magnitude (1 to 2 points) is only half that of the standard errors of measurement on the WAIS-R (i.e., 2 to 4 IQ points) (p. 967). Interesting hypotheses about possible generational changes or changes in tests have been put forward to explain these observations. However, it is perfect circular reasoning to first carefully construct a test to show minimal sex differences, and then to conclude from data collected with that test that there are only trivial sex differences. The large-scale metanalyses of abilities ought really to show no sex difference at all had the test constructor done his work properly. What is at stake is that test construction, and the interpretation is influenced more by ideology than by reality, and that data collection is corrupted by the sex-dimorphic approach. It is actually very easy to construct an intelligence test showing either female or male overall superiority, depending on the number of verbal and spatial items included. The model presented in the second part of this chapter represents an attempt to avoid the traps of sex dimorphism and ideology by taking its point of departure in different individuals rather than in average sex differences, and by accounting for sex-related variability without supporting any particular sexual ideology.

B. Theoretical Problems

1. Introduction We have a number of more or less mutually exclusive genetic, social learning, interaction, and cognitive theories to explain sex differences in abilities. This is a clear sign of confusion. In this section I first outline some of the theoretical restraints that give rise to bewilderment, and then suggest a reorientation that is less vulnerable.

2. Genetic Explanations Genetic explanations of specific abilities basically come in three versions: the X-linked recessive-gene hypothesis, the polygene hypothesis, and the multifactorial hypothesis. The X-linked recessive-gene model was originally proposed by O'Connor (1943) and later supported in studies of Stafford (1961, 1963, 1965). The model assumes that a "good" gene for high spatial ability exists with a certain frequency in a given population. The gene is recessive and is linked to the X chromosome. The two sexes have an equal chance of receiving the gene, but males with one X chromosome need receive only one "dose," and females with two X chromosomes need the gene in double "doses" to express high spatial ability. Leaving out details here, the model predicts various familial transmission-pattern frequencies for spatial ability. These predictions found some support in the early studies, but did not hold up in more recent studies (Bouchard & McGee, 1977; Corley, DeFries, Kuse, & Vandenberg, 1980; Nyborg & Nielsen, 1981a, 1981b; Vandenberg & Kuse, 1979). It is now generally acknowledged that the X-linked recessive-gene hypothesis for specific abilities is inadequate.

Polygene models posit that the expression of specific abilities depends on a number of separate genes, each with a small additive effect. Multifactorial models acknowledge that multiple genes and multiple environmental factors combine in the expression of abilities. A major problem with such models is that we still do not know which genes combine with precisely which environmental factors. Neither the causal mechanisms nor the levels at which the effects are mediated are identified. How can the effects of, say, sex stereotypic rearing be transmitted to the brain, and where in the brain do they meet to affect the expression of specific abilities? There is another perhaps more serious problem with genetic models. Theories of population genetics and behavior genetics are applicable at the level of populations only. As discussed later (and elsewhere, Nyborg, 1977; Nyborg & Sommerlund, 1992), this restriction makes it virtually impossible to identify causal factors at the level of the individual, and this is the information we need most. Heritability coefficients are of great value for determining the average impact of genetic transmission, but they reflect a high level of statistical abstraction of little relevance for understanding the individual. To identify the reasons for the recently observed dynamic perturbations of specific abilities in adult females we have to move to the level of the single individual.

3. Social Learning Theory   Undoubtedly, only few learning theorists would insist that social experience is the only determinant of the development of specific abilities, but most emphasize the profound effects of upbringing, of social norms, of stereotypes, and so forth on abilities, at the expense of biological factors. Females are often said to excel in verbal areas because they passively model female role figures, or because they are positively or negatively reinforced to do so. Sherman's (1967) "Bent-twig" hypothesis presumes an early biologically based inclination for superior female verbal ability, and that this "Bent" is later socially forged into the adult sex difference in verbal abilities. We like to do what we do best. Boys have an initial bent for spatial ability, so they foster this ability at the expense of verbal ability. Genetic differences are seen by many social learning theorists as nonexistent or quite irrelevant. In this they follow the lead from the early American version of behaviorism. They further see the brain as sexually undifferentiated or tend to downplay effects of known sex differences in neuroanatomy and brain function on abilities.

The primary reason why socialization theory failed to explain the development of sex differences in specific abilities lies in the fact that putative causal factors like upbringing, modeling, norms, and stereotypes have never been operationalized. Mechanisms for transcribing putative exogenous effects in the social field into "endogenous" norms or attitudes and then again into effects on the expression of specific abilities, are left to speculation (Nyborg, 1983, 1992a, 1992b, 1993a). The principles of social learning theory make intuitive sense, but the empirical basis is essentially nondirectional correlations. As correlations do not prove causation, exclusive reliance upon them precludes further progress with respect to elucidating causes, effects, and direction.

Whereas social learning theory tends to view the brain as equipotential, most neuropsychologists have little problem with accepting the possibility that there may be important sex differences in the way the brain is organized and functions. What, then, is the neuropsychological evidence for sex differences in the brain?

4. Neuropsychological Explanations

a. Absolute Brain Size   Studies of sex differences in the size of the brain and of size-specific ability relationships have not been particularly informative until quite recently. Many studies show that males have a larger brain than females, and this was originally taken as proof that males are more intelligent than females. It was then argued that females may actually have the same size or a relatively larger brain than males, with control for body weight. The notion of a direct brain size–intelligence relationship was under fire during this long-standing and sometimes emotionally colored dispute. Defenders of the brain size–intelligence connection referred to a well-documented relationship between brain size and "intelligence" across species. Critics pointed to the fact that some of the largest brains were found in idiots, and very small brains were observed in highly gifted people. However, various aspects of the techniques for measuring brain size were criticized, and so was the quality of the brain preparations examined.

Until recently it was therefore generally accepted that the sex difference in corrected overall brain size was modest at best, and that size per se probably did not mean much to level of abilities. Then Willerman, Schultz, Rutledge, and Bigler (1991) measured brain size with magnetic resonance imaging (MRI), and were able to correct for individual differences in skull thickness and body size. They found that the high-IQ group (WAIS-R full-scale IQ 130; $M = 136.4$, $SD = 3.95$) had larger brain average ($p < .05$) than had the average IQ group (WAIS-R full-scale IQ 103; $M = 90.5$, $SD = 8.12$). A sex difference in adjusted brain size–IQ correlation was not significant ($p = .08$) but large enough (female $r = .35$ vs. male $r = .65$) to suggest a possible sex difference in brain organization. Johnson (1991) used a method developed by Van Valen (1974) and found an estimated brain size–intelligence correlation of .29. Ankney (1992) reanalyzed data on sex differences in brain size from autopsy records of 1261 25–80-year-old subjects originally studied by Ho, Roessmann, Straumfjord, and Monroe (1980). The reanalysis indicated that the absolute brain mass of white males and females of average height differ by 135 g, and that correction for body size reduces the male lead to 100 g. A similar result was obtained in black female–male comparisons. After correction for stature and weight, Rushton (1992a) found a male lead in cranial capacity (i.e., 1442 versus 1332 cm³, as derived from external head measurements) in a stratified random sample of 6325 United States military personnel. Finally, Rushton (1992) calculated cranial capacity from external head measurements collected by Jurgens, Aune, and Pieper (1990). He noted a 173 cm³ male lead in a very large population of 25–45-year-old males and females.

Obviously, these findings will appear controversial to many. The measures, the formulas, and the interpretation undoubtedly will be carefully scrutinized by skeptics. However, the consistency in the data from independent studies presents us with some interesting puzzles. Males seem to have a larger brain than females. Brain size seems to correlate moderately with overall intelligence. Common intelligence tests have been "adjusted" not to show sex differences. Recent large-scale studies nevertheless suggest a slight male advantage in overall intelligence. It is, in fact, possible to formulate various sophisticated hypotheses to make ends meet. My suggestion is simple. The sex-dimorphic approach indiscriminately bundles sometimes very different individuals and prevailing ideology perverts test construction. The cure is to apply individual-centered approaches and to become aware of and counter the damaging effects of sexual ideology.

b. Brain Anatomy   More and more sex differences are found in the neuroanatomical organization of the brain. The right hemisphere may be thicker than the left in males, but not different in females (Diamond, Dowling, & Johnson, 1981). The area of planum temporale is larger in the left than in the right hemisphere, irrespective of sex (Witelson & Pallie, 1973), and relatively larger in females than in males (Wada, Clarke, & Hamm, 1975). The functional significance of this has been related to superior female verbal ability, but the evidence for this is not strong. Animal data (mostly from rats) have long testified to the existence of a sexually dimorphic nucleus in the preoptic area of the hypothalamus (Gorski, 1984; Jacobson, Davis, & Gorski, 1985; Tarttelin & Gorski, 1988), and to sex differences in the corpus callosum, a broad band of fibers that connects the two hemispheres transversally (Denenberg, Berrebi, & Roslyn, 1988). These differences are moderated by exposure to gonadal hormones. There is some confusion as to the functional significance of this anatomical sex difference in the rat brain. It may relate to the regulation of male sexual behavior, although previous sexual experiences of the animal apparently also play a role (Jonge et al., 1989). Recently Swaab and Fliers (1985) reported a similar finding in humans, namely, a nucleus in the preoptic area of the hypothalamus that differentiates according to sex. The differentiation begins at postnatal age 2 when the female nucleus loses cells at a dramatic rate. At age 50 the rate of cell loss again accelerates, but this time in males as well as in females (Hofman & Swaab, 1989; Swaab & Hofman, 1988). The implications of this for sexual behavior in old age or in general are not known, and this particular sex difference has not been demonstrated to relate to specific abilities. A controversial topic is whether there is a human sex difference in the size of the corpus callosum. An early report on a larger female than male corpus callosum (Lacoste-Utamsing & Holloway, 1982) triggered speculations about whether this anatomical sex difference could explain the presumed more bilateral female brain organization and better verbal ability. A number of subsequent studies were unable to confirm the original observation of a sex difference in the corpus callosum, or gave only partial support (O'Kusky et al., 1988), using various measurement techniques, including MRI and cerebral blood-flow approaches. A recent study suggested that there probably is no sex difference in overall anatomical size of the corpus callosum, but there may be a sex difference in its shape (Allen, Richey, Chai, & Gorski, 1991).

Excellent reviews of these and many other sex differences in the brain anatomy can be found in Goy et al., 1980; Kimura, 1983, 1987; McGlone, 1980, 1986; and Swaab, Hofman, and Fisser, 1988. The caveat is that most of these data are collected within the constraints of the sex dimorphic approach, which maintains that genetic, genital, or self-reported sex is a befitting classification variable.

c. Functional Brain Organization   The early studies of sex differences in functional brain organization yielded a mountain of puzzling data, and their interpretation is problematic. We know, for example, that a simple functional left–right brain terminology with related absolute sex differences in verbal-performance IQ needs qualification. Early lesion studies indicated that damage to some areas in the right side of the brain lowers spatial abilities more in males than in females, but tends to leave male verbal abilities relatively unaffected, whereas damage to the left side of the brain lowers verbal abilities more in males than in females, but tends to leave male spatial abilities relatively intact. These observations led to the assumption that the male brain was functionally organized more unilaterally, and the female brain organized more bilaterally. It has gradually transpired, however, that degree of functional brain lateralization, kind of ability, level of task complexity, handedness, and sex, are all interrelated. Many attempts have been made to identify the subserving neuroanatomical substrates, and to determine the developmental timetable for the lateralization processes, but newer evidence suggests that the developmental brain architectural–functional relationships are more complex than a simple left–right sex-dichotomous terminology would allow for. It may very well be that both male and female brains are slightly differently lateralized for some relatively automated executive motor functions, and that circumscribed areas in the female left anterior hemisphere may be more lateralized or better organized for verbal skills than the female posterior and the male anterior and posterior left hemisphere (Kimura, 1983; Kimura & Harshman, 1984; McGlone, 1980). Moreover, with increasing task complexity both verbal and spatial performance seem to call upon concomitant neural activity in both hemispheres, although perhaps to a different extent in the two sexes.

Further adding to complexity in the area, Juraska (1984, 1986; Juraska, Fitch, Henderson, & Rivers, 1985) and others have demonstrated sex differences in the effects of experience on neuroanatomical development. Harris (1980) and Nyborg (1983) have, among others, called attention to the possibility of anatomical and/or functional sex differences in subcortical areas with an effect on patterns of selective activation or inhibition of cortical activities.

A relatively new line of research—psychoneuroendocrinology—suggests that it might be a good idea to call upon a more molecular approach in order to elucidate the functional aspects of brain–ability relationships. This is so, because steroids obviously hold exciting potentials for explaining concomitant development of the brain and abilities. Prenatal and later brain effects of hormones may, in fact, constitute one of the most direct ways to experimentally combine structural with functional aspects of brain neuroanatomy and chemistry. To give an example, early steroid treatment affects subcortical development and ability. Pavlides, Westlind-Danielsson, Nyborg, and McEwen (1991) treated neonatal rats with thyroid hormone, a treatment which speeds up prepubertal hypothalamic development and initially improves learning performance compared to untreated controls (Schapiro, Salas, & Vukovich, 1970). However, adult rats treated neonatally with thyroid hormone demonstrated defective spatial learning in an eight-arm radial maze, and also showed impaired electrophysiological long-term potentiation in the dentate gyrus of the hippocampus. This was taken to reflect reduced synaptic efficacy and neural plasticity. Other steroid influences will be outlined later in connection with a presentation of psychoneuroendocrinological evidence for sex-related differences in the developmental pattern of the left and right cerebral hemispheres and for different ability patterns in subgroups of homosexual men.

5. Psychometrics Truly quantitative approaches, such as psychometrics, have yielded an invaluable body of descriptive data on individual and sex differences in specific mental abilities. Unfortunately, even the most exact psychometric study runs into problems when it comes to explanations. Some doubt was recently expressed on empirical grounds, whether purely descriptive nonexperimental approaches—such as psychometrics—will ever disclose the nature of putative brain processes subserving specific abilities (Lynn, 1990; Nyborg & Sommerlund, 1992; Vernon, 1990). The only way to disclose whether knowledge about neuroanatomy and functioning applies to specific abilities is through the study of the brains of single individuals (Nyborg, 1977), and we urgently need a coherent theoretical framework that does not confuse different levels of explanation (Nyborg, 1993a).

C. Concluding Remarks

Genetic, social learning, neuropsychological, and psychometric studies have provided invaluable information about average sex differences in specific abilities. The major problem with these studies is, however, that they do not provide us with the information we would like to have most of all. We want to know exactly what are the cause(s) of sex differences in specific abilities. We want to know details of their causal nature, their mechanisms, and their biological locus of action. We want to identify the proper level of description of interactions among the various causal agents involved. We want to establish a precise, comprehensive, and testable theoretical framework in order to better understand relevant observations made in other fields. None of the traditional theories offer that much. The genes, held responsible for sex differences in abilities, have yet to be identified. Too much social learning theory is based on anecdotal evidence, intuition, correlation coefficients, and many results or interpretations are carefully adjusted to conform to sexual norms for equality or for political correctness. Most neuropsychologists operate with empirical data, and some even rise to the level of experimental control, but neither clinically oriented nor experimentally oriented neuropsychologists have yet been able to formulate a comprehensive, coherent, and generally acceptable neuropsychological theory about the brain-specific ability connections. Psychometrics is strong on descriptive analysis and on quantification, but is in principle and in practice unable to outline the brain mechanisms behind specific abilities.

I suspect that the main problems with traditional theories is that they all accepted at face value the sex-dimorphic strategy. Sex dimorphism is not just a misleading term. It inevitably confounds within-sex high- and low-ability males and females (see also Nottebohm, 1980). This indiscriminate within-sex mixing of sometimes fundamentally different individuals, superficially made similar by sexual classification, may be the most serious obstacle on the way towards proper operationalization of cause(s) and mechanisms. With the possible exception of neuropsychology, most theories further operate simultaneously at several incompatible levels of explanation. Social, genetic, mental, and neural levels are tied together by unspecified interaction processes. These two objections go, I believe, to the heart of reasons for the conspicuous lack of progress in the study of the causal nature of sex differences in specific abilities. We need a fresh approach!

A fresh approach must, first of all, take its point of departure in the study of single individuals (Nyborg, 1987a). The individualized approach enables us to probe directly into those parts of the brain that are believed to be functionally important for the appearance of differences in specific abilities, while at the same time this approach fully appreciates the large individual variation. Only if enough individuals were found to be similar with respect to a given characteristic would we have stumbled over something genuinely general (Nyborg, 1977; see also Allport, 1962; and Runyan, 1983). The within-group variation reflects the extent to which it makes sense to talk about a modal trait. This is the opposite of trying to establish valid principles for individuals on the basis of statistical averages or factors. With respect to sex differences, the study of modal trait development allows for the obvious possibility that some traits show a sex-*related* tendency, but in that case we have to operationalize sex more adequately than phenotypic or self-reported sex allow for (see Section 8). The fresh approach should further enable us to directly address questions about proximate causes for individual and sex-related differences in brain development, and to identify specific mechanisms and loci of biological action. The new approach should allow us to operate at one level of explanation instead of many. This is the way to escape the problem of incompatibility. The approach should be empirically testable at each step of exploration and explanation. The new approach should allow us to empirically examine details of concomitant ontogenetic brain development and associated individual and sex-related differences in specific abilities. The new approach should, finally, facilitate the interpretation of the results within a phylogenetically meaningful perspective of ultimate causes. Obviously, all this means nothing less than a paradigmatic shift in the traditional exploration of sex differences in specific abilities. This shift would imply a rather reserved look at data obtained through sex-dichotomous procedures, and a keen interest in safe identification of the cause(s) for covariant individual and sex-related differences in body, brain, and ability development.

III. PART II: THE GENERAL TRAIT COVARIANCE–ANDROGEN/ESTROGEN MODEL

A. Introduction

The general trait covariance–androgen/estrogen (GTC–A/E) model reflects an attempt to formalize effects of genes, gonadal, hormones, and experience, on body, brain, and behavior in a way that overcomes problems associated with the sex-dimorphic approach (Nyborg, 1979, 1983, 1988a, 1988b; Nyborg & Nielsen, 1977). The model represents, in fact, an attempt to explain complex individual and sex-related covariant trait development as a combined function of the reciprocal effects of genes, hormones, and experience on body and brain tissues. The model presumes that males and females are made individual by genes, experience, and hormones, and that useful predictions can be derived after hormotyping them. A complete hormotyping also incorporates family dispositions (heritability estimate of genetic transmission until more precise DNA measures become generally available) and personal life history events (including prenatal experiences), but most of this remains a task for the future. As soon as an individual is defined according to hormotype, the model generates predictions about the most likely development of his or her body, brain, and behavioral trait pattern in a form that can be tested immediately. The present neuropsychological context makes it natural to emphasize the molecular gene–environment–hormone–brain–behavior mechanisms used to explain unfolding of the specific ability patterns. The model is presented graphically in Figure 1. Besides specific abilities this version of the model also deals with a few of the other traits that tend to harmonize with the particular ability pattern of the different hormotypes. A more elaborated version of the model dealing with more traits is presented elsewhere (Nyborg, 1992a, 1992b), as is evidence that the model successfully predicts covariant personality trait development (Nyborg, 1984). The theoretical framework behind the model is called physicology, and is presented in part in Nyborg (1988a, 1989, 1992a), and in detail in Nyborg (1993a).

Only a few of the many molecular mechanisms believed to mediate steroid-based development and function will be outlined here. The field is vast, and grows at an exponential rate, so the discussion will be focused on mechanisms by which gonadal hormones regulate gene expression prenatally and later in life, and on the recently observed capability of gonadal hormones to quickly change nerve membrane characteristics and thus mimic effects of neurotransmitters. Neither do space restrictions allow a detailed overview of hor-

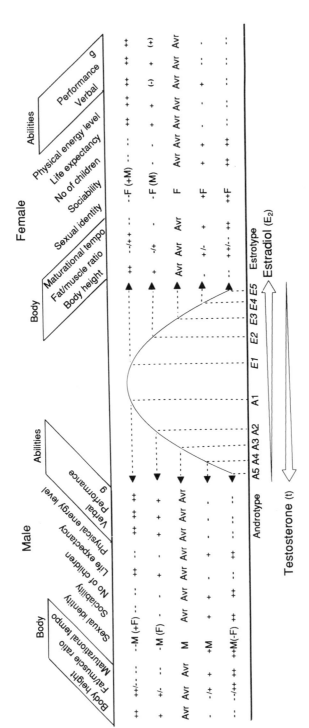

FIGURE 1

The General Trait Covariance–Androgen/Estrogen (GTC–A/E) model for development. Males are classified in accordance with plasma testosterone (t) concentration into hormotypes A1 to A5, where A1 is low t and A5 is high t individuals. Females are classified in accordance with plasma estradiol (E_2) concentration into hormotypes E1 to E5, where E1 is low E_2 and E5 is high E_2 individuals. The GTC–A/E model generates predictions for the various hormotypes with respect to coordinated somatic, psychological, and behavioral trait development (see text for details). The present version of the model is borrowed in part from Nyborg, 1979; 1983; 1984; 1987b; 1988a; 1990b; Nyborg & Nielsen, 1981b.

monal effects on neuronal architecture, but good discussions can be found in Toran-Allerand (1984, 1986). As mentioned previously, the GTC–A/E model readily acknowledges that gonadal hormones do not work in isolation. Various experiences transiently or permanently affect the production of gonadal hormones, as for example under short-term or long-term stress. The causal pathway for experience goes, according to physicology, through physico-chemical changes in the peripheral sensory systems as a function of variation in the entirely physical environment (including physical people's physical behavior). The causal pathway also includes hormonally mediated modulation of gene transcription and neurotransmitter changes in the brain, often with an effect on molecular memory systems, and sometimes with an effect on the neuroregulation of executive motor systems leading to overt behavior. Nonetheless, it is constantly under the influence of positive or negative feedback and feed forward adjustment from various physicochemical systems. In this respect the model represents an attempt to explain the extremely complex and dynamic molecular aspects of interactions among experience, gonadal hormones, genes, and the brain.

B. The Dual Role of Hormones in Evolution and the Ontogenesis of Specific Abilities

1. Evolutionary Aspects   A brief look at evolutionary theory helps illustrate why physicology considers steroids the most central of ultimate causes in the evolution of life forms with complex central nervous systems and so-called intelligence. The following section outlines why physicology also puts steroids, together with DNA material and experience, in the role as *the* proximate causal agents in the ontogenesis of individual and sex-related differences in specific abilities.

Natural and sexual selection is, according to Darwin, the driving force behind evolution. Selection takes place among different individuals, so it became essential for Darwin to identify the origin of individual differences. Unfortunately, Darwin never opened a letter he received from a then little-known Austrian monk, Gregor Mendel, which explained the rules for heritability of simple traits. Darwin was therefore forced to reluctantly resort to a kind of Lamarkian thinking, which acknowledges that the environment exerts a heritable impact on the individual. However, when Watson and Crick broke the genetic code it became obvious that robust chemical instructions, in the form of particular combinations of four relatively simple nucleic acids, explain stable species-specific ontogenetic development, and that structural and numerical mutations plus gene combinations plus environment superimpose ontogenetic individualization on otherwise remarkably stable species-specific programs.

Gonadal hormones are vital for the evolution of complex life forms in general (Hapgood, 1979; Roth et al., 1982; Symond, 1979; Witzmann,

1981). With respect to specific abilities, there are several good reasons why gonadal hormones become of particular interest (Nyborg, 1992a, 1993a). First, they enable the evolution of the sexual reproductive mode, and this confers a major competitive advantage with respect to superior sexual selection: It became possible to quickly and efficiently combine favorable mutations and new gene constellations for body and brain structures and functions that best subserve survival and reproduction of parents and of the next generation. This is one reason why gonadal hormones became an important ultimate evolutionary cause for successful adaptation to life-threatening challenges in a variable world. Second, besides conferring an evolutionary edge through the appearance of the sexual reproductive mode, gonadal hormones have another long-term advantage. During evolution they exerted profound and very direct effects on the timing of the single individual's ontogenesis, that is, on the unfolding of various anatomical and functional body and brain structures, in close coordination with environmental requirements. The most obvious evolutionary advantage of this is a survival bonus to species that responded fast and adaptively to harsh conditions during upbringing and in later phases of life. In this way reproductive capability could be adjusted effectively with nutritional and stressing conditions to save precious energy. To make a long story short, gonadal hormones constitute, according to the physicological viewpoint, the molecular bridge in the evolution of complex life forms between accumulated effects of numerical and structural mutations and of gene recombination, the adaptive regulation of body and brain development and reproductive behavior, and sexually biased social interaction. To the extent that gonadal hormones and reproductive aspects color so-called social interaction, gonadal hormones constitute the proximate basis upon which all complex forms of society have to be understood, according to physicology. They provide the molecular basis for Freud's speculation that sex penetrates all aspects of culture. The point is that species are certainly made in accordance with specific genetic information in DNA, but the gradual making of maleness and femaleness (or of everything in between) during evolution presumes proximate and rather specific effects of gonadal hormones regulating protein production of the genome in hormophilic tissues. Without gonadal hormones there will be no sex-related differentiation, and the evolution of species with complex central nervous systems would have been slower or even missing completely.

2. Ontogenetic Aspects   For a long time it was believed that there are two phylogenetically stable gene programs for sexual differentiation: one for female development and one for male development. The male program was believed to be the female program modified by additional information from the Y chromosome. Femaleness was, in other words, seen as more basic, and maleness as a kind of add-on to the female program. Recent findings indicate that all fetuses more likely begin body and brainwise as sexually undifferentiated, except for the presence of a Y chromosome in male fetuses. The Y chromosome provides for the formation of testicular tissues, and the later androgen output down-regulates genes for female development, and up-regulates genes for male development. It has further been suggested that prenatal female brain development depends on relatively low estradiol (E_2) output, whereas higher output of E_2 masculinizes the brain (Döhler & Hancke, 1978; Döhler et al., 1984). Whether this can be verified remains to be seen, but it is safely established that androgen can be aromatized to E_2 in the brain, and that gonadal hormones regulate genes to promote male and female variations in initially neutral body and brain tissues (McEwen, 1987a, 1988a), in reproductive behavior (Gorski, 1974; McEwen, Jones, & Pfaff, 1987; Romano, Mobbs, Lauber, Howells, & Pfaff, 1990), and in nonreproductive behavior at the most complex levels (Hoyenga & Hoyenga, 1979; Nyborg, 1979). This newer experimental evidence indicates that if there are some kind of phylogenetically stable gene programs for separate phenotypic male or female differentiation, they can be entirely overruled by manipulation of gonadal hormone content during early phases of development, by blocking actions of relevant hormones, or by influencing the induction of specific receptor proteins for gonadal hormones.

Early in the century it was realized that infrahuman and other animal reproductive behavior depends on the presence of gonads and, by inference, on hormones. With improvements in techniques for refining and measuring gonadal hormones it gradually became obvious that not only human body and brain development, but also performance remotely related to reproductive functions, was profoundly affected by hormones. The expression of verbal and spatial abilities, as well as of personality was seen to be significantly influenced by naturally occurring and medically provoked hormone perturbations. These observations line up with the hypothesis that all sex-related differences in body, brain, and behavior are a function of proximate effects of gonadal hormones, and that the sexually most dimorphic traits show the strongest sex hormone involvement (Hoyenga & Hoyenga, 1979). Talking about effects of gonadal hormones it is worth noting that both sexes produce "male" and "female" sex hormones, and that "male" hormones often act as prohormones for "female" hormones in both sexes. This makes terms like "male," "female," and "sex hormone" rather ambiguous. They will, therefore, in the following be substituted with the more neutral expression "gonadal hormone," or be referred to by their family name (e.g., androgen and estrogen), or by their specific name (e.g., testosterone, t, and estradiol, E_2, to mention some of the more important hormones in the present context.

C. Theoretical Implications

The choice of gonadal hormones as *the* ultimate and proximate causes for sex-related differentiation of specific abilities carries with it a number of theoretical implications. Being an integrated part of the physicological framework, they are better made explicit here before a presentation of the GTC–A/E model. Otherwise, the new paradigm might too easily be confused with elements of the traditional sex-dimorphic view or we might continue to operate at several levels of explanation. For example, it makes absolutely no sense within a physicological framework to assume that gonadal hormones affect abstract mental entities. Gonadal hormones become biologically active exclusively through molecular interactions with genes, other chemicals, or membrane proteins, and this can have an effect on body and brain architecture and function. The GTC–A/E model is designed in accordance with this, and does not operate with concepts like "mind," "mental" or "cognitive" ability, "thinking," or "general intelligence." These abstractions have no place in a modern molecular account of behavior. They may, however, be used (with extreme caution) as convenient shorthand descriptors, as long as it is clearly understood that they refer to complex 100% material processes going on in the brain. Mental concepts are, in other words, completely stripped in physicology for the explanatory value previously ascribed to them in the mentalist framework. They have no causal status whatsoever. This means that hormones are not assumed in the GTC–A/E model to guide mental development, and they have neither "purpose" nor "goal." Gonadal hormones rather increase or decrease the probability that body and brain development takes particular directions, by regulating the expression of species-specific genes of importance for the appearance of particular traits. When reference is made in the following to the development of a positive manifold of abilities, such as for example, the sum of verbal and nonverbal abilities, Spearman's (1904) more neutral designator, g, is used, again under the assumption that we mean brain function and not mind.

Another aspect of the new paradigm is that the level of explanation is confined to one and one only, namely, the molecular (Nyborg, 1988b; 1992a; Nyborg & Bøggild, 1989). The emphasis on the material basis of sex-related differences in specific brain abilities makes it possible to provide coherent explanations of their development. There actually may be no other scientifically acceptable way to avoid the classical, anthropocentric, dualistic body–mind trap, but this is not the place to discuss details of the arguments for the physiological point of view, for which the reader is referred to Nyborg (1993a). Suffice it here to state that the choice of gonadal hormones as the proximate agents for sex-related differentiation makes it possible to circumvent most of the problems associated with traditional sex dimorphism. Classification of individuals in accordance with their hormotype (see Section D) solves another problem. Males typically have higher plasma androgen concentrations and lower E_2 than females, but the levels of gonadal hormone in plasma are clearly continuously distributed, and there is overlap between male and female distributions. This encourages a nondichotomous classification of individuals. Add to this that "male" hormones often act as prohormones for "female" hormones in both sexes, and that the rate of conversion from one hormone to another varies on an individual basis, and the advantage of using models that operate with continua on the causal as well as on the effect side becomes obvious. It follows that the tradition of dividing individuals into two sexes is better substituted by the more flexible and individualized classification, and that the expression "sex-related difference" is, if not perfect, a more adequate term than "sex difference." The only reason for not exclusively using the term "individual differences" is that quite often a particular modal pattern of specific abilities actually goes together with self-reported sex. However, the ability pattern of some individuals can be very different from that of their same-sex fellows, and closer to that of the heterotypic sex (depending on genes, experience, and hormotype). In such cases "sex differences" becomes a misleading term.

The new paradigm necessitates a revision of our classical nature–nurture model. This is so, because gonadal hormones regulate the protein production of the genome throughout life, and this violates several basic assumptions of the traditional nature–nurture interaction model. Some implications of this are discussed later.

After these brief remarks about the dual role of gonadal hormones in evolution and ontogeny, I turn to the task of hormotyping. Procedures for classifying individuals are presented first, and then the phenomenon of trait harmonization is alluded to. Examples of trait harmonization are provided with the presentation to illustrate the role of hormones in this orchestration of trait development.

D. Hormotyping

There are two basic ways to classify individuals according to hormotype: (1) indirect estimates through anthropometry, behavioral evaluation, or questionnaires, and (2) direct measures of gonadal hormones in plasma, urine, or saliva.

1. Indirect Estimation   Indirect estimates of hormone status can be made on the basis of quantitative or qualitative examination of various objective body characteristics (Marshall & Tanner, 1986) that typically differ among the sexes. Among those used are shoulder and hip width, amount and distribution of fatty tissues, muscular strength, distribution of hair, and deepness of voice (Bourguignon, 1988; Crockett & Petersen, 1985; Farthing, Mattel, Edwards, & Dawson, 1982; Petersen, 1976; Petersen & Taylor, 1980; Waber, 1976, 1977a, 1977b, 1979; Young & Reeve, 1980). Such traits are known to be affected by gonadal hormones (Farthing et al., 1982; Tanner, 1975). Other indirect indicators of hormonal status are aggression and dominance, or self-confidence and sexual (gender) identity, often measured by questionnaires, such as the Bem (1983) Scale or Berzins, Welling, and Wetter's (1978)

Personality Research Form (PRF) Androgyny Scale. The PRF Andro Scale was, for example, used successfully to show that individuals with high masculinity or femininity score (assumed to reflect high androgen or E_2 status, respectively) encountered problems in nonverbal ability tests, whereas androgynous individuals with some high masculinity *and* some high femininity score (assumed to have intermediate hormone values) as well as sexually undifferentiated individuals with low male *and* low female score (assumed to have low hormone status) did well in tests for spatial ability, irrespective of genetic sex (Bøggild & Nyborg, 1992). Berenbaum and Resnick (1982), McKeever (1986), and Petersen (1976), found that physically androgynous males and females score higher on spatial ability tests than do sexually more developed individuals.

Despite some success, neither anthropometric measures nor scores on personality questionnaires can be said to provide very accurate estimates of hormonal status (Knussman & Sperwien, 1988). There are several reasons for this. Individuals differ in sensitivity to circulating hormones, and the sensitivity varies over time, because of changes in the number of hormone receptors induced, which relates to changes in the maturational state of tissues. In addition to gonadal hormones, adrenal hormones also affect body and brain development, and many other factors contribute to biochemical variation. This suggests that indirect estimates of hormotype should be used with caution, and only when direct hormone determination is out of the question.

2. Direct Measures Techniques for direct measurement of plasma gonadal hormones have improved considerably over time, but there are still problems. Blood sampling is a stressing event to some, and stress affects the endocrine system. One sampling of blood is the minimum, but a series of consecutive blood samples is preferred because of the pulsatile nature of hormone secretion. Particular individuals may refuse to participate in studies requiring blood sampling. Laboratories often use different techniques or chemicals for assaying, which means that data from one laboratory cannot safely be compared with hormone data from another laboratory without a common quality control. Hormone content is related to age, weight, medication, nutrition, and race, and techniques for analysis undergo constant improvement, which complicates the evaluation of longitudinal measures. Hormone content can further be estimated by analyzing urine, typically collected over a 24-hr period. Urine sampling is less stressing than collecting blood, and allows compensation for diurnal variations in hormones that are secreted in brief pulses followed by long intervening periods with little secretion. A drawback is that the estimation of plasma hormone concentrations from urine typically is based on measurement of metabolites of the plasma hormones. It has become possible to measure some gonadal hormones in saliva (Landman, Sanford, Howland, Dawes, & Pritchard, 1976; Read, Riad-Fahmy, Walker, & Griffiths, 1982; Sannikka, Terho, Suominen, & Santti, 1983). Sampling saliva is not particularly stressing, and saliva provides a good measure of the biologically free fraction of t. However, most other hormones cannot yet be determined with sufficient accuracy in saliva. Thus, care has to be taken when estimating an individual's hormotype, irrespective of whether direct or indirect methods are used.

3. Classification Young females typically have 3–5 times higher E_2 than males, and can be ordered linearly according to plasma E_2, a biologically very active member of the estrogen group. The E_2 distribution can, for example, be a priori divided in five arbitrarily chosen intervals, and the intervals are named E5 to E1. E5 and E1 in that case represent females with the 20% highest and lowest E_2 values, respectively, and E3 refers to the 20% females with close to average values.

Males can be hormotyped in the following way. They are linearly ordered according to individual t (a biologically very active member of the androgen group). The t distribution is a priori divided into five arbitrarily chosen intervals named A5 to A1, where A5 and A1 represent males with the 20% highest and lowest t values, respectively, and A3 represents the 20% individuals with close to average values. Young males typically have much higher plasma t concentration than young females. In addition to the problems mentioned with hormotyping females, the hormotyping of males has its own problems, however. For reasons given elsewhere (Nyborg, 1979, 1994b) the GTC–A/E model assumes that E_2 is the most important steroid in the brain for understanding hormone-ability relationships. Moreover, E_2 may feminize the brain in low concentrations and masculinize it in high concentration (Döhler & Hancke, 1978). So, the body may essentially be feminized by E_2 and masculinized by t or their metabolites, but for the brain E_2 might be the more important hormone. It complicates matters that females aromatize some of their brain and peripheral t to E_2, but males do so to a larger extent. Obviously, an evaluation of the effects of a particular hormone on the brain is made difficult when the hormone measured in plasma, say t, might masculinize the body and at the same time actually act as a prohormone for E_2 with effects on the brain that may differ according to dose. There is now a number of animal study groups looking into these problems. They use, among other techniques, radioactive labeling of gonadal hormones and follow the pattern of uptake and the biological and behavioral effects with rather sophisticated techniques. This chapter is not the place to review the evidence, and the reader is referred to a brief discussion of how the GTC–A/E model handles limited aspects of this complex problem in Nyborg (1983). To illustrate general principles of male adult hormotyping it suffices here to mention that as boys pass through the various pubertal phases, they begin to differ markedly with respect to plasma t concentration. A complete hormotyping also includes evidence for familial dispositions. A probabilistic estimate of what genes can be expected to be modulated by hormones is called for, as is exact information about possible permanent prenatal hormone effects on body and brain development. There are many possibilities. Atypical fetal hormone secretion, variations in the mother's hormone status due, for example, to stress or medication, or even the hormone secretion of a cotwin (in particular if the cotwin is male) may affect the fetus. Major socioeconomic and life-history events should also be recorded in connection with hormotyping. There are not many studies of the heritability of hormone status, but those at hand suggest that the hormotype is under considerable genetic control. Meikle, Stringham, Bishop, and West (1986) thus found that about 40% of the variability in t in males can be explained by a familial disposition. In an earlier study of monozygotic and dizygotic twins, Meikle, Bishop, Stringham, and West (1986) found that genes regulate between 25% to 76% of the total variation of plasma t, E_2, estrone, 3α-diol G, free t, luteinizing hormone (LH), follicle-stimulating hormone (FSH), but only 12% in dihydrotestosterone (DHT) and <1% in sex-hormone-binding-globulin (SHBG). Moreover, the male hormotype seems to remain fairly stable over time, as repeated measures of t show short-term stability (Couwenbergs, Knussman, & Christiansen, 1986; Knussman & Christiansen, 1986) and remains fairly constant over a year (Smals, Kloppenborg, & Benraad, 1976). However, there are systematic circadian and circannual changes in plasma t content (Nieschlag, 1974; Reinberg, & Lagoguey, 1978; Reinberg, Lagoguey, Chauffournier, & Cesselin, 1975), which should be taken into account when sampling. The female hormotype also seems to remain reasonably stable over the reproductive period, as repeated measures of plasma E_2 tend to give similar readings from one E_2 high phase to another and from one E_2 low phase to another over consecutive menstrual cycles.

Obviously, the proposed hormotyping represents the simplest possible solution to an immensely complex problem. It is assumed, for example, that female plasma E_2 content does not correlate with plasma t content, and that male plasma t does not correlate with E_2 content. However, preliminary analyses suggest that E_2 and t do correlate significantly in 8 and 10 yr old girls ($r = .57$ and .69, respectively) but not in boys of similar age, and that the correlation remains positive but not significant thereof (Nyborg, 1994c). No doubt further research will lead to more sophisticated ways to classify individuals hormonally, but in complex matters it sometimes pays off to first see if simple solutions work. Genetic, genital and bodily sex, and sexual identity typically (but certainly not always) coincide with a predominant t or E_2 exposure. Effects of extreme hormone exposure are discussed at length by Hoyenga and Hoyenga (1979); Nyborg (1984, 1990a); and Nyborg and Nielsen (1977, 1981a, 1981b).

4. Trait Covariance   As said before, the primary goal of the GTC–A/E model is to account for harmonization of traits during development. Gonadal hormones are ideal for this purpose, because they go everywhere in the body.

Unlike many other substances in plasma, gonadal hormones easily transcend the blood–brain barrier, so they can affect the body and the brain at one and the same time. On the other side, gonadal hormones become biologically active only in target tissues that are able to induce hormone-specific receptors. This arrangement puts gonadal hormones in a unique position. Despite their simultaneous availability in all body and brain tissues, they nevertheless can exert precise and highly circumscribed effects only in designated target areas. The task of the GTC–A/E model is to formalize the developmental and functional implications of this arrangement. By simultaneously and selectively influencing body and brain target tissues, gonadal hormones become capable of precise coordination of widespread but circumscribed aspects of body and brain development and functioning. Variations in gonadal hormones may force single bodily or behavioral traits to appear, disappear, or reappear in the phenotype, or change intensity, frequency, or direction of the expression of traits. More importantly, they concomitantly orchestrate the various aspects of sex-related body, brain, and behavioral development, including specific abilities, and may thus profoundly modify ontogenetic timetables for development. The GTC–A/E model is devised to explain why various hormotypes develop different combinations of body, brain, and behavioral traits, why pubertal tempo differs among hormotypes, why different hormotypes exhibit different degrees of adult sexual differentiation of the body and the brain, and thus why specific abilities and personality come to differ among hormotypes in adulthood.

E. Model Predictions

Figure 1 is a graphical representation of the GTC–A/E model. The present version is restricted to explain only individual development of specific abilities and a few other characteristics. It will be remembered that hormotypes A3 and E3 represent individuals with close to average male t values or average female E_2 values, respectively, and average sexual differentiation. A4 and E4 designate a male with somewhat higher than average t, or a female with somewhat higher than average E_2 concentrations, both showing a higher degree of sexual differentiation. A2 and E2 mirror a male or a female with somewhat lower than average t or E_2, respectively. Hormotype A2 and E2 approach so-called androgynous males and females, that is, males who in addition to ordinary masculine traits also show some clearly feminine traits, or females who in addition to the usual feminine traits also show some clearly masculine attributes. There are problems with proper definition of the dimensions of masculinity, femininity, and of androgyny (e.g., Lenney, 1979; Spence & Helmreich, 1979), but this is not the place to take up that discussion.

The GTC–A/E model generates person-specific testable predictions about the most likely trait combination as soon as an individual is classified according to hormotype. An individual with a moderate plasma concentration for his or her sex like a male A2 or a female E2 is expected to be tall and show either a fatty (A2) or a leptosome (E2) body build. They are expected to be late maturers, to have a late sexual debut, and to display an androgynous sexual identity. There might be a slightly increased tendency for bi- or homosexuality in A2 males (with low prenatal t concentration, or during the first few years after birth; Swaab & Hofman, 1988), and for E2 females with high prenatal t (Ellis & Ames, 1987; Ellis, Peckham, Ames, & Burke, 1988; Feder, 1984; Money, Schwartz, & Lewis, 1984; Swaab, Hofman, & Fisser, 1988; Whitam & Mathy, 1991). This, however, is probably far from the rule, and needs further investigation (Gooren, 1988). A2s and E2s are not very sociable. They tend to be loners and prefer books and work over people. They marry late, and typically get one or few well-in-advance planned-for children. They live longer than average for their respective sex. In addition to these similarities, there are also predictable differences among A2s and E2s. A2 males are expected to have been less physically aggressive during upbringing than most boys, whereas E2 females are expected to have behaved physically more aggressive than usual for girls. A2s and E2s may, nevertheless, be rather high in verbal aggressiveness and score high on persistence. With respect to specific abilities, both A2s and E2s are expected to score higher than average, after correction for family dispositions, of course. They typically get high performance scores, but E2s are expected to get a slightly lower than average verbal score than A2s, for reasons to be discussed.

A4s and E4s with higher than average homotypic plasma gonadal hormone concentrations for their sex are expected to show a developmental pattern that in many respects represents the diametrical opposite of that of the A2s and E2s. A4s and E4s tend to be short, stocky, or athletic early-maturing individuals. A4 males show extensive bodily masculinization and E4 females display a pronounced feminization of the body. Sexual identity is unanimously and stereotypically in line with their respective sex. A4s and E4s are expected to prefer extensive social activities at the expense of book reading, to marry early, and to have many children. They may show shorter life expectancy than A2s and E2s, perhaps due to bad health consequences of high plasma hormone content (Ellis & Nyborg, 1992). A4 males are expected to display an increased level of physical aggression, relative to A3s, but female E4s most likely display more submissive behavior than E3 females (Baucom, Besch, & Callahan, 1985). With respect to abilities, A4s and E4s are expected to show overall below-average g, and significantly lower than that of the A2s and E2s. Although A4s and E4s may excel in some of the more automated aspects of verbal skills, such as verbal fluency, and in motor skills, relative to E3s and A3s, they are expected to display a severe depression of information and nonmotor performance subscale score. This, obviously, detracts from overall g, which is defined as the sum of verbal and performance scores.

Socioeconomic background is often claimed to determine achievement and vocational status, and to explain why low g individuals tend to cluster in low-status occupations. An alternative explanation is that g codetermines achievement in certain occupations (Harrell & Harrell, 1945; Jensen, 1980; McCall, 1977; Terman, 1925; Waller, 1971), obviously, together with other factors. To the extent that specific abilities are actually called upon in certain high-status jobs, the GTC–A/E model predicts that the lonely, relatively infertile (Vining, 1984) androgynous A2 and E2 individuals are more likely to be found near the top of the educational, occupational, and political pyramid (Editorial, 1976; Hingley & Cooper, 1983). The model further predicts that the more socially inclined A4 and E4 individuals will excel in areas requiring physical power, aggression, or dominance rather than high g (Dabbs & Morris, 1990; Dabbs, de La Rue, & Williams, 1990; and section A.5).

F. Mechanisms

The literature on mechanisms through which gonadal hormones exert their many and varied effects on body and brain target tissues expands rapidly. Excellent overviews are provided by, among others, Döhler et al., 1984; Goy et al., 1980; McEwen, 1987b; Nottebohm, 1981, 1989; Nottebohm, Nottebohm, and Crane, 1986; Toran-Allerand, 1984; and other chapters in the book edited by Vries, Bruin, Uylings, and Corner (1984). The literature suggests that gonadal hormones have profound effects on cell proliferation, migration, and differentiation, as well as on the functionality of mature cell assemblies. The general causal pathway seems to be as follows. After secretion, gonadal hormones are carried around in the bloodstream. In this phase most hormones are biologically inactive by being associated to binding proteins, SHBG, with such high affinity that most steroids appear in plasma in a bound form, leaving only a few percent active. Figure 2 (from Norman & Litwack, 1987) illustrates the mechanisms of action of hormones on the genome.

The figure shows that unbound hormone molecules enter the target cells and are bound there to unoccupied receptors (2) to form an inactivated complex (3). It is not clear whether this happens in the cytoplasm or in the nucleus, so two pathways are indicated in the figure. The steroid-receptor complex is accordingly either activated while still in the cytoplasm (4) and translocated to the nucleus (5), or activated while in the nucleus (4). In either case the steroid-receptor complex exhibits an increased affinity for DNA (6), and binds to presumptive high-affinity binding sites upstream from the genes (6), with the result that the gene is regulated by the hormone. As the transcription rate of the gene increases, the thus accomplished mRNA is translocated (7) to the cytoplasm for translation. The final result of steroid action on the gene is, in other words, that the protein production of particular genes is modulated. The enhanced level of the new proteins can then be utilized for multiple purposes, such as to build up specific body and brain tissues during development, to induce neurotransmitters or influence their working condi-

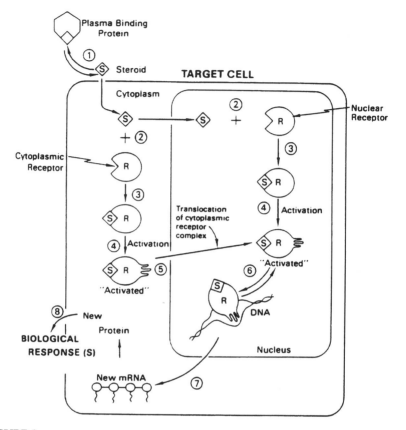

FIGURE 2

Diagram of how plasma sex hormones regulate the expression of genes. Most of the sex hormone (S) is chemically bound by protein in plasma (1), but the free fraction can enter the target cell (2) and is bound to a receptor (R), that is either in the cytoplasm or in the nucleus (3). After activation (4) the *cytoplasmic* hormone-receptor complex enters the nucleus (5) and then binds (6) to a part of the gene with high affinity. Alternatively, the *nuclear* hormone-receptor complex is activated (4) and then binds directly (6) to the gene. There is some discussion as to where the intracellular hormone receptor is situated, and accordingly which way the receptor-hormone complex goes, but in either case the result is gene transcription, new messenger RNA, and a new protein that can be utilized in body and brain tissues for structural or functional ends. (Reproduced with permission from Norman & Litwack, 1987, p. 23.)

tions, or to alter the metabolism of various remote tissues and organs with further multiple effects for other organic systems. Understanding of the details of these intricate processes is still incomplete. Do receptors for gonadal hormones reside in the cytoplasm or in the cell nucleus in the absence of hormones (Toran-Allerand, 1984)?

Another question is whether gonadal hormone work exclusively through the regulation of gene products, or affect each other and neurotransmitters? Gonadal hormones sometimes have an effect with such a short time delay that they could not possibly work exclusively through the rather time-consuming actions on the genome (McEwen, 1988a, 1991b). It appears that some steroids and their metabolites are capable of modulating the release of neuroactive substances, chloride flux, and the distribution of oxytocin receptors in the hypothalamus via membrane receptors within a time span of minutes (McEwen, 1991a, 1991b). Moss and Dudley (1984) have illustrated how E_2 may affect nerve cells in a manner similar to that proposed for certain neurotransmitters by a membrane-mediated effect (see Figure 3).

Estrogen molecules can interact with receptors on the cell membrane to stimulate the intracellular production of cyclic adenosine monophosphate (cAMP), which then binds to an intracellular receptor and affects other metabolic processes in the nerve cell. According to this model, E_2 needs not actually enter the cell during this process, and bypasses the time-consuming translocation to the nucleus, the subsequent transcription of gene proteins, and the alteration in protein synthesis, but results in a change in cell metab-

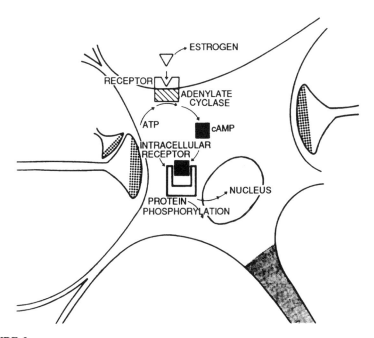

FIGURE 3

Diagrammatic representation of a membrane-mediated mechanism of estrogenic action, which is similar to that proposed for neurotransmitter action. ATP, adenosine triphosphate; cAMP, cyclic adenosine monophosphate. (From Moss and Dudley, 1984.)

olism. These details of mechanisms are interesting by themselves, but do not affect the general argument made here. It is sufficient for an understanding of the GTC–A/E model to know that gonadal hormones profoundly affect the anatomy and functionality of numerous bodily and brain tissues in various ways. They can permanently alter the neuronal architecture and function. They can change the organization or secretional pattern of glands remotely situated from where the chain of molecular events began. The reactions may take place within seconds, hours, days, or months. The fact that gonadal hormones initiate cascades of local effects as well as guide more remote molecular interactions, explains how even minor perturbations in gonadal hormones are capable of profoundly influencing the orchestration of rather different body, brain, and behavioral functions, depending, among other things, on the maturational status of the organism. Moreover, these cascades of causal effects often unfold in nonlinear ways.

Another important point in steroid chemistry is that gonadal hormones often show almost opposite effects in low and high doses. E_2 exerts strong neurotropic effects even in tiny doses (Toran-Allerand, 1984), but it may have neurotoxic effects in high or very high concentrations. The possibility of circumscribed or diffuse neurotoxic reactions to long-term medication with steroids (Nyborg & Nielsen, 1981b), or to constantly high physiological concentrations (Ellis & Nyborg, 1992) should not be underestimated. The concept of natural cell death has become increasingly popular in recent years (Bowen, 1981; Hinchliffe, 1981). It is based on the observation that the organism produces many more cells during development than is needed, so an important question is which mechanisms accomplish the systematic eradication of surplus cells. Steroids seem to play an important role in programmed cell death (Gould, Woolley, & McEwen, 1991; Greenwald & Martinez-Arias, 1984; McEwen, 1988a, 1991a; Truman, 1985; Truman & Schwartz, 1984).

Until a few years ago it was customary to divide effects of gonadal hormones into two groups: the early organizational and permanent neuroarchitectural effects on immature body and brain target tissues, and the late activational effects, transiently influencing the functioning of mature body and brain tissues. However, recent animal evidence indicates that the adult brain of some bird species remains sufficiently plastic to undergo major anatomical seasonal changes (DeVoogd & Nottebohm, 1981; Nottebohm, 1981, 1989; Nottebohm, Nottebohm, & Crane, 1986). The adult rat brain responds to hormone variations over the estrous cycle with dendritic changes and alteration in the number of synaptic connections (Woolley, Gould, Frankfurt, & McEwen, 1990). Considering that the human EEG pattern changes dynamically over the menstruating cycle (Becker, Creutzfelt, Schwibbe, & Wuttke, 1980; Klaiber, Broverman, Vogel, & Kobayashi, 1974), as a function of the female brain being exposed to varying concentrations of gonadal hormones, and that the use of oral contraceptives has a stabilizing effect on the human EEG (Matsumoto, Sato, Ito, & Matsuoka, 1966), it may very well also be that the adult human brain responds with microscopic neuroanatomical or molecular changes to natural or provoked hormone perturbations.

IV. NEUROPSYCHOLOGICAL APPLICATION OF THE GTC–A/E MODEL

A. Early Structural Effects

Gonadal hormones appear to condition differences in the development of the left and right cerebral hemispheres. Irrespective of sex, the right hemisphere grows larger than the left in several species, including the human (Kolb, Sutherland, Nonneman, & Whishaw, 1982), and it has been suggested that t is responsible (Diamond et al., 1981; Geschwind & Galaburda, 1987; Steward & Kolb, 1988). However, the male right hemisphere becomes slightly thicker than the female (Diamond, 1988). The functional significance of this anatomical asymmetry for specific abilities is not transparent, and the sex-dimorphic nature of the studies may explain mixed results in the area. We already know of early significant asymmetries in the number of hormone receptors in the two brain halves (Sandhu, Cook, & Diamond, 1986) and also subcortically, and that receptors come and go with age. Moreover, there are complex region, sex, and developmental asymmetries in those brain-aromatization processes by which t is metabolized to E_2 (Ziegler & Lichtensteiger, 1992). We now need data from large-scale studies of hormonally different individuals rather than from sexually classified groups to learn about details of the origin, developmental significance, and proper meaning of the observations for specific abilities.

B. Developmental Dynamics

1. Prepubertal Development   As stressed previously, it is essential to keep a developmental perspective on the evaluation of differences in specific abilities. It is generally acknowledged that most sex-related differences in abilities do not appear, or typically are of moderate size, before puberty (Maccoby & Jacklin, 1978). One exception is the early and persistent female lead in some verbal areas. With respect to spatial abilities, a few studies even report a slight prepubertal female lead. However, these observations may be explained by the generally more advanced prepubertal female body and brain development.

2. Pubertal Development   Ability patterns change at puberty according to sex-dimorphic studies, but these studies often raise more questions than they answer. One unanswered question is whether boys continue to improve in spatial abilities during late puberty, whereas girls' performance reaches an early asymptote, or whether boys' spatial abilities reach an asymptote at puberty and girls' spatial abilities actually decline. It was concluded in an earlier review (Nyborg, 1983) that many girls (and some boys) actually regress in spatial abilities, but ideology sometimes prevents a proper interpretation of this phenomenon (Witkin et al., 1967). The GTC–A/E model allows us to formulate specific and testable hypotheses to answer the above question. Recall that the model predicts low spatial abilities in females with high-E_2 plasma content, and high spatial abilities in females with low E_2 content. Let us for a moment use the model as a tool for examining developmental dynamics, and exploit its curvilinear dose–response characteristics. First we map female biological age along the X-axis in Figure 1, in parallel with the one for individual age-related variation in plasma E_2 concentration. Girls enter puberty in accordance with a gradual increase of plasma E_2, but there are large individual differences in the size of the E_2 surge. These differences can be used for prediction. The moderate increase in E_2 in hormotype E2 girls partly explains why she matures late and why she might show signs of virilization. She does not "overshoot" the optimum brain E_2 range by much, so the prediction is that her family disposition for spatial ability will be only slightly depressed, if at all. A girl with more ample age-related increase in plasma E_2, such as hormotype E3, matures at the average female age, and displays average female sexual differentiation. She "overshoots" the optimum brain E_2 level more than an E2 girl, so the prediction is that she suffers more depression of familial disposition for spatial abilities than an E2 girl. A girl with a considerable surge in plasma E_2 at puberty, that is hormotype E4, is expected to mature at an early age and to show a high degree of secondary sexual feminization. She "overshoots" the optimum range by a wide margin. The prediction is that highly feminized E4 girls suffer the most severe depression of spatial abilities. The model further predicts that the depression in spatial abilities takes place at different chronological ages in the various hormotypes. E4s are expected to show the earliest depression and E2 the latest and least, with E3 in between. For girls in general the regression seems to take place some time between ages 12 and 14 (Nyborg, 1983, 1988a, 1990a; Nyborg & Nielsen, 1977), but we need studies aimed specifically at testing these predictions.

The GTC–A/E model generates, in other words, testable predictions about who regresses in spatial abilities, when, why, and how much. The basis for this is quite simple. Examine covariant hormonal, bodily, brain, and ability development in terms of the effects of hormones, and keep the analysis on a person-specific basis. It then becomes obvious that the highly feminized E4 girl runs a particular risk of early and severe depression of nonverbal abilities, that the moderately feminized E3 girl suffers less and later depression, and that the androgynous E2 girl shows little or no depression of spatial abilities. The traditional sex-dichotomous approach averages the scores of these three groups, and thereby blurs the picture. Obviously, nothing in the GTC–A/E model speaks against an even finer differentiation of individuals into more than five hormotypes. Such fine-grained analyses are best made in forth-coming specific and uncompromising testing of the above-mentioned predictions, preferably in the form of rigorous longitudinal or cohort-sequential designs (e.g., Nyborg, 1994c).

Low plasma t prepubertal boys grossly "undershoot" the optimum brain E_2 level because there is little brain t to aromatize to E_2. As they approach puberty, some boys (hormotype A2s) exhibit only a moderate increase in t, delayed body and brain maturation, and an androgynous body type and sexual identity. A2 boys "undershoot" the brain E_2 optimum, and the low t gives more allowance for E_2 to compete for receptors sensitive to both hormones. The result is almost full expression of spatial abilities in A2s. Boys with an average surge in plasma t at puberty (A3s) "overshoot" the brain E_2 optimum due to aromatization, but not nearly as much as the E3s. They suffer slight depression of spatial abilities. A3s display average masculinization of the body and sexual identity. A4 boys with a considerable plasma t surge at puberty "overshoot" the brain E_2 optimum by a wide margin, suffer a significant depression of spatial ability, become highly masculinized with respect to body characteristics, and display a stereotypic male sexual identity.

These examples suggest that classification of individuals according to hormotype gives room for meaningful differentiation among individuals, prenatally, at puberty, and later in life, whereas the sex-dichotomous approach confounds this important within-sex differentiation. The hormotypic approach suggests that when males and females are low in androgens and estrogens, respectively, they also differ least in somatic sexual differentiation and in their specific ability pattern. Conversely, when males and females are high in plasma gonadal hormones they also differ most in sexual differentiation of the body and abilities.

C. Menstrual Dynamics

The model predicts inverse changes in specific abilities as a function of cyclic changes in E_2. Remember that E3 and, in particular, E4 females are assumed to "overshoot" the optimum brain E_2 by a wide margin with resulting pubertal depression of the spatial abilities and enhancement of the verbal abilities (see Figure 4, from Nyborg, 1979; modified, 1992).

We can exploit the fact that the GTC–A/E model sees the expression of specific abilities in adulthood as a dynamic rather than as a static phenomenon as in sex-dimorphic models. The model predicts that if plasma and brain E_2 concentration changes in an adult female, the balance between the spatial and verbal abilities will be affected in a quite specific way. They must change in perfectly opposite directions, or the model would be in error.

Dynamic effects of menstrual variation in plasma E_2 on spatial abilities have, in fact, been observed in several independent studies (Anderson, 1972; Dor-Shav, 1976; Hughes, 1983; Klaiber, Broverman, Vogel & Kobayashi,

1974; Komnenich, Land, Dickey, & Stone, 1978; Silverman & Phillips, 1991). Woodfield (1984) administered the Embedded Figures Test to pregnant women and again after childbirth. Spatial ability was low in the last high-E_2 weeks before birth and increased in the weeks after, when E_2 returned to normal values. Progesterone, which varies slightly out of phase with E_2, is probably not responsible for the hormone effects on abilities (Hampson, 1990). Gordon, Corbin, and Lee (1986) intravenously administered LH-releasing hormone (LHRH) and LH, two feminizing hormones, to males and found that scores for verbal production went up, whereas mental rotation spatial scores decreased.

Some studies are unable, however, to find a consistent effect of cyclic gonadal hormones on specific abilities (Gordon et al., 1986; Ho, Gilger, & Brink, 1986; Woodfield, Whitehead, & Asso, 1987). Moreover, positive studies do not always find a hormone effect in equivalent spatial tests. Small sample size, inaccurate timing of hormone sampling, measurement error, and large within-sex variability probably all contribute to the partly inconsistent picture. Moreover, the dynamic effects of sex hormones on specific abilities may not be of dramatic proportions, even though Hampson (1988) found that 18% of the individual variability in the Paper Folding Test could be explained by absolute differences in plasma E_2. Obviously, further research is needed to reveal the precise nature of cyclic hormone effects on spatial abilities.

The apparent noncyclic nature of male hormones may camouflage that

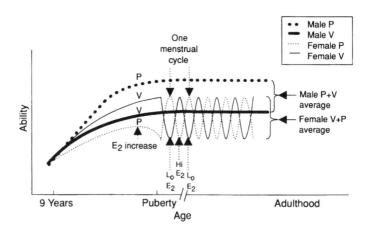

FIGURE 4

Sex differences in adolescent unfolding of verbal (V) and performance (P) abilities. At puberty male V and P abilities reach an asymptote, whereas female P is inhibited. During the female reproductive period, V (and fine-motor skills) will cycle in opposite phase with P (and with visual-spatial skills), as a function of effects of cyclic menstrual changes in the sex hormone estradiol (E_2) on the brain. Lo E_2 = Perimenstrually low plasma estradiol (E_2) concentration. Hi E_2 = Midcycle high plasma estradiol (E_2) concentration. (Nyborg, 1979, 1992.)

gonadal hormone perturbation also affects specific abilities in males. Kimura (1991) thus found that male spatial ability is higher in the spring, when t tends to be low, than in the autumn when t is higher.

D. Adult Male Abilities

I plotted best-fitting polynomial curves for a number of intelligence test scores as a function of plasma t in 3654 middle-aged white American males (raw data from Centers for Disease Control, 1988, 1989) (Nyborg, 1994a). Males with higher than average t (i.e., ≥ 678 ng/dL) got lower general technical scores the higher the t on the Army Classification Battery (ACB) and lower WAIS Information subtest, and WAIS Block Design scores. Males with below average t got higher scores on the same tests the lower their t. Scores on the General Technical Aptitude Test and number of years of formal schooling also followed this pattern. Log t level was significantly inversely related to age in this large sample (Ellis & Nyborg, 1992). The data further suggested that the major races differ with respect to the influence of variation in t concentration on abilities, but are equally sensitive to the effect of high t on depression (Nyborg, 1992). Also, Shute, Pellegrino, Hubert, and Reynolds (1983) found that spatial abilities were best in low t males and high t females, but there might have been problems with the antibody used for t determination (see Gouchie & Kimura, 1990). Gouchie and Kimura also noted better spatial and mathematical abilities in low t males and in high t females than in the reverse, but they were unable to see an effect of t on tests that usually favor females or show no sex difference.

Christiansen and Knussman (1987) found that t correlates positively with measures for spatial abilities, and negatively with verbal abilities, and Tan and Akgün (1992) noted a positive correlation between t and scores on Cattell spatial reasoning in a small sample of young right-hand, right-eye-dominant males, but mixed-dominant males and young females showed no relationship. These observations suggest that it is important to determine exactly where on the curvilinear position of the GTC–A/E model the subjects under examination are located.

Undoubtedly, detailed information about genetic dispositions for abilities and laterality, agreement about which test to use for comparison, uniform assaying techniques, and a protocol for when to sample blood, urine, or saliva, would greatly help to clarify details of the somewhat mixed pattern of the t–ability relationship. Guidelines for exact values of where to cut gonadal hormone distributions into meaningful hormotypic intervals would also be useful. Based on the axiom that all sexually differentiated traits show proximate hormonal involvement, and more so the more differentiated they are, it further becomes of interest to determine which modal "male" and "female" traits will cycle as a function of changes in t and E_2 in addition to spatial, verbal, fine-motor and gross-motor skills. In general, traits that require only pubertal hormone activation can be expected to cycle. Traits that require both prenatal hormone priming and later pubertal activation can also be expected to cycle. Traits for which prenatal hormone exposure is sufficient to appear may not cycle in adulthood. Our present knowledge of which traits need which kind of hormonal action to be fully expressed is insufficient.

E. Educational and Occupational Aspects

As mentioned previously, specific abilities hold moderate power for predicting achievement in educational and occupational areas. Does hormotyping fare any better in this respect? The question is really whether the hormotype, with its associated covariant body, brain, ability, and personality development, constitutes a more reliable platform for predictions than do abilities or genetic sex? Indirect support for this notion would be if hormotypes appear with different frequencies in particular educational and occupational areas. In fact they do. Purifoy and Koopmans (1979) found that professional women have more plasma t than housewives. In the language of the GTC–A/E model this means that a late-maturing, androgynous, high-ability E2 female enjoys an advantage in occupational achievement, whereas the early-maturing feminized social E4 female cares more for children than for a professional career. The model would by implication, at least partially, explain the fact that g differs among occupations (see Section A.5). Schindler (1979) observed higher t in female attorneys than in athletes, nurses, or teachers. The observation of low t in female athletes should come as no surprise because hard physical work depresses t. Dabbs et al. (1991) classified several smaller and one large population of 4462 middle-aged males (same database as used by Nyborg, (1992, 1994a; and by Ellis & Nyborg, 1992) in accordance with occupational criteria, and then looked for differences in t. Ministers had the lowest, and blue-collar workers, football players, and actors the highest plasma t concentrations. It even seems that specialization within a given occupational area is related to differences in plasma t, and that there is an inverse relationship between t and socioeconomic status. Dalton (1976, 1979) found that girls exposed in utero to progestin (often with androgenlike effects) had more academic success than had untreated control sisters. Girls exposed to more than normal prenatal t tend to do better in math and science than their nonexposed sisters (for an early overview, see Hoyenga & Hoyenga, 1979).

We need more research to fully explore connections between hormotype, abilities, and educational and occupational achievement. The advantage of drawing upon models like the GTC–A/E model for this task is essentially that the hormotype and associated trait combinations offer a precise platform for testing predictions of achievement in particular educational or occupational areas. Ability scores and genetic sex by themselves are not enough.

F. Homosexuality

The precise cause(s) for homosexuality is(are) not known, so explanations range from exclusive genetic determination, over abnormal hormone conditions in the prenatal (but not in the adult) period (e.g., Ellis & Ames, 1987), to exclusively social determination. Homosexuality does not ordinarily figure in neuropsychological literature, and certainly not in connection with a discussion of specific abilities. There might be good reasons for making the connection, however, because of its potential for revealing details of connections between prenatal hormone exposure, brain development, and the unfolding of specific abilities. McCormick and Witelson (1991) thus inquired into the neurobiological etiology of homosexuality and specific abilities. They reasoned that if male homosexuality is due to lower than normal t exposure during the prenatal period, and further if handedness and specific abilities also reflect differences in prenatal t exposure, then there might be differences in ability patterns among left- and right-handed male homosexuals, as compared to matched male and female heterosexual controls. They found that the mean performance of homosexual males on three spatial tests and one verbal fluency test fell between those of the heterosexual males and females, but was not significantly different from that of the control females. In this they corroborated the findings of others. They further noted significantly more not-consistently right-hand users in the male homosexual group who, like other such handed individuals, also showed a tendency to obtain lower spatial relative to verbal fluency scores. McCormick and Witelson explain these observations in terms of an effect of low male prenatal t on brain organization, abilities, and handedness. Female fetuses, exposed to abnormally high levels of t (e.g., girls with the adrenogenital syndrome or through medication) later display an increased probability of left-handedness (e.g., Nass et al., 1987), a "male" pattern of high spatial ability (e.g., Resnick, Berenbaum, Gottesman, & Bouchard, 1986), and an increased probability of a bisexual or homosexual inclination (Ehrhardt et al., 1985; Money, 1987). Females with abnormally increased prenatal t exposure would probably qualify as hormotype E2 or E1, and males with abnormally low prenatal exposure would count as hormotype A2 or A1. These hormonal variations over a broad physiological range suggest interesting ways to exploit predictions of the GTC–A/E model with respect to hormonally mediated neuronal links between sexual orientation, laterality, and specific abilities.

G. Evolution

Section B.1 briefly outlines two good evolutionary reasons for selection of gonadal hormones during evolution: the advent of the sexual reproductive mode, and the adaptive ontogenetic timing of sexual maturation as a means to conserve energy by not producing much offspring during hard times. There is a third good reason for selection of hormones in connection with selection for specific abilities. The general form of this argument is, briefly, that males and females became selected for different abilities early in evolution in accordance with a practical division of labor. Males were selected for good spatial abilities on wide-range hunting expeditions, whereas females became selected for traits like sociability, empathy, and verbal abilities. The trait combinations of a dominant, muscular, quiet male hunter with excellent spatial and gross-motor skills, and a subordinate, nutritive, sociable, and talkative child-breeding female was, according to this view, a highly competitive and well-matched arrangement that would be strongly selected for. Views differ, however, about at which level selective forces took effect. Some think they worked at the level of the single gene, others that they worked at the level of specific behavioral traits including specific abilities, or at the level of the individual, or at the level of the population, or at the level of society or culture. Sociobiologists speak of gene–culture coevolution. Not many researchers consider evolution a matter of pure physics, or assume that selective forces can have only an entirely physical target (Atkins, 1981; P. M. Churchland, 1984; P. Churchland, 1986; Jacob, 1982a, 1982b; Monod, 1975). Physicology is an attempt to apply the physical point of view in behavioral fields like psychology. Physicology is defined as the physicochemical study of very complex carbon-based so-called living systems (Nyborg, 1993a). Physicology maintains that it is highly unlikely that physical selective forces can have an effect on mental entities or that behavior as such can be a target for physical forces. Behavior is conceived of as an intermediate product of a series of complex endogenous physicochemical processes interacting constantly with exogenous physicochemical processes as long as so-called life processes take place. Aggression or spatial ability as such cannot, according to this view, be selected for during evolution, but the molecular constellations that give rise to such traits and to their combinations can. The point is, that selective forces in the environment, and this includes other people, are of a purely physical nature, and so are the developmental factors and processes responsible for the unfolding of specific brain-based abilities and the body. Physicology substitutes the psyche with movement of molecules. This brief discourse returns us to the suggested classification of individuals in accordance with their hormotype. Early organisms lacking steroids were doomed to depend on primitive predominantly nonsexual modes of reproduction, which are not particularly beneficial with respect to fast and efficient combinations of genes favorable for the evolution of complex brains and related abilities. Then, some kind of primordial steroid precursors appeared on the evolutionary scene, most likely as by-products of a series of accidental mutations. The path was then set for selection to eventually favor organisms capable of composing a primitive egg full of nutrition and later primitive egg-and-sperm arrangements. The capability to reproduce partly or fully sexually, gives a decisive competitive edge. In other words, the physicochemical basis for the evolution of complex brains and their associated abilities is, according to physicology, repeated unplanned DNA mutations and later selection for steroids that regulate the expression of the DNA material. It is generally acknowledged that evolutionary processes stabilize at some optimum level, depending in part on the selective characteristics of the ecological niche. The physicological explanation of the origin and relative stabilization of today's prevalent trait combinations, represented by the various hormotypes, is that they reflect cost-benefit rudiments of ancient selection processes. Moderately low-hormone E2 and A2 individuals have superior abilities, but they get few children (Vining, 1982, 1984), and successful reproduction and survival of fertile offspring is the final evolutionary test for the continuation of a species. On the other side, low abilities in high-hormone E4 and A4 individuals is "compensated" for by an edge in high reproduction rate. Depending on the nature and changes of the ecological conditions in a particular evolutionary niche, one or the other hormotype will be selectively favored. From a neuropsychological perspective, this means that the particular central nervous organization and the associated ability pattern of the hormotypic variations of modern Homo sapiens sapiens reflects selective pressures of the past. One of the implications of this is troubling. The high-reproductive–low-ability trait constellation in A4 and E4 individuals may have served well and was favored in primitive times, when physical force and surplus children were necessary prerequisites for the survival of the species. Unfortunately, these very same characteristics seem to be dysgenic in a modern world, where reproductive restraints and good abilities are required. Under such ecological circumstances, A2 and E2 individuals will be favored, educationally, occupationally, and otherwise, and this leaves us with a serious problem.

H. The Nature–Nurture Question

We now know that gonadal hormones regulate genes throughout the life span. The regulation takes a dramatic form prenatally, when the female or male developmental program is switched on by hormones. This regulation raises, in fact, serious doubt about the adequacy of the traditional nature–nurture model, which is based on Fisher's analysis of variance model. Genetic and environmental components are assumed to each make an independent contribution, and together they add up linearly to a 100% phenotypic appearance. Less impact of one factor automatically means more of the other. Such models cannot accommodate for the fact that a change in the environment may induce a change in gonadal hormones, which may permanently activate or deactivate parts of the genome (McEwen, 1988c, 1988d; Nyborg, 1989), but this happens time and again. It happens prenatally if the hormonal milieu of the fetus is affected by medication of the mother. It also happens when the mother becomes severely stressed, or as a function of maternal or placental hormone perturbations, or even via the hormonal contribution of a cotwin. Environmentally mediated changes of the expression of the genome take place after birth when stimulation of the peripheral sensory system and the brain affects the endocrine and other systems. In each of these cases environmental factors can and do accomplish a permanent change in the genetic contribution to ontogenetic development. Finally, besides linear and additive effects, hormones probably also exert nonlinear multiplicative effects. The traditional nature–nurture model is not at all geared for all these possibilities, and this raises an urgent need for revision for hormonal (Nyborg, 1989, 1990b) as well as for other reasons (Wahlsten, 1990).

I. Race Differences in Abilities

There is in the sex-dichotomous literature fair agreement about the size of race differences in abilities, but little agreement about the reasons for the differences. Some argue that tradition, cultural factors, discrimination, victimization, or lack of learning opportunities explain the low spatial abilities observed in many Blacks, the relatively high spatial abilities seen in many Asians, and the intermediate spatial abilities typically observed in Whites. Others argue in favor of a genetic or a gene–environment interaction explanation. The physiological approach differs from these explanations on two accounts. First, it requires a precise operationalization of relevant causal factors and mechanisms. Second, it requires an examination of individuals and their hormotype rather than remaining satisfied with averages. With respect to causal factors there is evidence that young Black males have significantly more t than White males (Ross et al., 1986), and with respect to individual differences it is equally clear that there are large within-race differences in t (Ellis & Nyborg, 1992). Less is known about Asians, but they are believed to have lower hormone levels than others (Purifoy, 1981). In terms of hormotyping we can make a very preliminary classification: Many Black males will qualify as A4s. Many White males will be A3s. Many Asian males will be A2s (Nyborg, 1987b). Given these assumptions, the GTC–A/E model predicts that many Black males will "overshoot" the optimum brain E_2 range, due to aromatization of surplus t, with a resulting marked depression of spatial abilities. Many White males will display a slight depression of spatial abilities, but most Asian males will show little or no pubertal depression. Blacks are expected to show the highest degree of sexual differentiation, Asians the least, with Whites in between. Race differences in verbal fluency and motor skills can be expected to show the exact opposite order, that is, higher verbal than spatial ability in Black males, slightly lower verbal than spatial ability in White males, and significantly lower verbal than spatial ability in Asians. There is some evidence in favor of some of these predictions, but the picture is far from clear. There are several reasons for this. The custom of averaging abilities for a whole race causes problems similar to those of averaging according to sex. Moreover, besides effects of actual person-specific environmental conditions (e.g., severe protein deficiency) on hormone status, patterns of inheritance also ought to be taken into account. Geographically determined effects of living for extended periods of time in particular ecological niches most likely play an important role for the gene pool (Nyborg, 1987b, 1994b). Points worth keeping in mind when considering these predictions are that within-race differences are larger than between-race differences, and also that the within-race differences are partially explained in terms of hormotype. Instead of continuing to look for significant average race differences we should perhaps rather compare the frequencies of certain hormotypes within and between races and use formal models to explain what we see in both instances.

V. GENERAL DISCUSSION

The main thesis of this chapter is that the traditional sex-dichotomous study of sex differences in specific abilities is beset with serious problems, and that it is time to introduce new paradigms. The one suggested here provides methodological and theoretical alternatives, and the question then becomes whether the new approach solves all the traditional problems. To answer this question I will briefly review each of the old problems in the light of the proposed solutions.

Does hormotyping actually solve the basic problem with the sex-dichotomous approach? The answer is essentially yes. Hormotype classification is made on the basis of an identified proximate cause, the value of which is distributed along a continuous dimension. The fact that the model for cause–effect relationships is curvilinear does not affect the general argument. We can classify any population into as narrow intervals on the hormone scales as deemed necessary. Neither is the positive answer affected by the fact that minor changes in gonadal hormones can lead to truly tremendous nonlinear cascadelike effects along the causal pathways towards the guidance of development and function of remote tissues. We actually seem to have found a way around the sex-dichotomous Procrustean bedstead. Hormotyping may partly solve another long-standing controversy in psychology: the clinical single case versus population average approach, or the nomothetic versus the idiographic approach (Allport, 1965; Cronbach, 1975; Eysenck, 1976). Male and female averages typically characterize only a few individuals within each of the two groups, and the large overlap in sex distributions makes it difficult to interpret the mean difference. Implicit in the use of sex averages is further the presumption that most males have in principle similar patterns of abilities, and so have females. However, high-and low-ability scores tend to outbalance each other within sex, large variability reduces the probability of finding statistically significant differences, and averages lead to loss of valuable information about individuals. The solution to this problem is to ban the almost exclusive use of the individual-differences approach, and to promote the use of the different-individuals approach, if we want to look for causes. Sex hormones give rise to continua in effects, and the observation of enough similar individuals may signal the possibility of a common biological basis. To be sure, the individual-differences and the different-individuals approach need not be mutually exclusive at the descriptive level. An average may signal that something general may be lurking behind, but the causal status of that something can be explored only in the study of individuals. The only exception to this rule would be the unlikely case that everybody is identical. Hormotyping provides a more individualized basis for predictions about educational and occupational achievement than sex- or average-specific abilities, because gonadal hormones are predisposed for individual harmonization of bodily, brain, and behavioral development. Several studies, even though not using hormotyping, have already demonstrated the value of taking hormones into account as predictors of education and occupation. This line of research can be expected to develop at a fast rate in the near future. Puberty is the time when sex-related differences in hormones and abilities really show up. However, research indicates that the prenatal period is very important for understanding these differences, and that also other phases of life characterized by hormone perturbations are relevant. The critical attitude towards research on race differences is to some extent based on the argument that within-group differences typically are larger than between-group differences and that causes for individual differences may not explain group differences. The hormotypic approach strives to explain the large within-group differences on an individual basis, while at the same time acknowledging that the frequency of certain genes and hormotypes may differ among racial groups. It is, in brief, argued that the focus on causal agents and on the individual saves hormotyping from much of the critique that deservedly is raised against the sex-dichotomous approach and against the exclusively genetic study of race differences.

It is only fair to state that the GTC–A/E model is nothing more than a preliminary heuristic tool for circumventing traditional problems in the neuropsychological study of individual differences in body, brain, and specific abilities. However, the model has its positive sides. It suggests links between ontogenetic development and evolution and facilitates the understanding of primordial sources (or ultimate causes) of contemporary sexual differentiation of specific abilities. The GTC–A/E model reveals shortcomings in Fisher's classical variance model for analyzing nature–nurture interactions during ontogenetic differentiation of specific abilities and suggests improvements. The model incorporates rapid advances in neuroendocrinology in the reformulation of flexible and person-specific models for nature–nurture interaction. The model can explain recent observations of dynamic effects of sex hormones on abilities, as well as covariant body, brain, and specific-ability development. The model is potentially falsifiable at each step of exploration and explanation. I believe that these characteristics will be minimal requirements for any new model striving to associate modern brain research with developmental neuropsychological research on individual and sex-related differences in specific abilities. The GTC–A/E model and the framework of physicology might provide the key to developing better paradigms in the field.

REFERENCES

Allen, L. S., Richey, M. F., Chai, Y. M., & Gorski, R. A. (1991). Sex differences in the corpus callosum of the living human being. *The Journal of Neuroscience, 11*, 933–942.

Allport, G. (1962). The general and the unique in psychological science. *Journal of Personality, 30*, 405–422.

Allport, G. (1964). *Pattern and growth in personality*. London: Holt, Rinehart and Winston.

Anderson, E. (1972). Cognitive performance and mood change as they relate to menstrual cycle and estrogen level. *Dissertation Abstracts, 33*, 1758-B.

Ankney, C. D. (1992). Sex differences in relative brain size: The mismeasure of women too? *Intelligence, 16*, 329–336.

Atkins, P. (1981). *The Creation*. San Francisco: W. H. Freeman & Company.

Baucom, D., Besch, P., & Callahan, S. (1985). Relation between testosterone concentration, sex role identity, and personality among females. *Journal of Personality and Social Psychology, 48*, 1218–1226.

Becker, D., Creutzfeldt, O., Schwibbe, M., & Wuttke, W. (1980). Electrophysiological and psychological changes induced by steroid hormones in men and women. *Acta Psychiatrica Belgica, 80*(5), 674–697.

Bem, S. (1983). Constructing a theory of the triple typology: Some (second) thoughts on nomothetic and idiographic approaches to personality. *Journal of Personality, 51*, 566–577.

Benbow, C. (1988). Sex differences in mathematical reasoning ability in intellectually talented preadolescents: Their nature, effects, and possible causes. *Behavioral and Brain Sciences, 11*, 169–232.

Berenbaum, S., & Resnick, S. (1982). Somatic androgyny and cognitive abilities. *Developmental Psychology, 18*, 418–423.

Berzins, J., Welling, M., & Wetter, R. (1978). A new measure of psychological androgyny based on the Personality Research Form. *Journal of Consulting and Clinical Psychology, 46*, 126–138.

Bøggild, C., & Nyborg, H. (1992). The development of spatial abilities: Activity, personality, and sexual identity. Unpublished manuscript.

Bouchard, T., & McGee, M. (1977). Sex differences in human spatial ability: Not an X-linked recessive gene effects. *Social Biology, 24*, 332–335.

Bourguignon, J. (1988). Linear growth as a function of age at onset of puberty and sex steroid dosage: Therapeutic implications. *Endocrine Reviews, 9*, 467–488.

Bowen, I. (1981). Techniques for demonstrating cell death. In I. Bowen & R. Lockshin (Eds.), *Cell death in biology and pathology*. London: Chapman and Hall.

Centers for Disease Control (1988). Health status of Vietnam veterans. I: Psychosocial characteristics. *Journal of the American Medical Association, 259*, 2701–2707.

Centers for Disease Control (1989). *Health status of Vietnam Veterans, Vol. I: Synopsis*. Atlanta: Centers for Disease Control.

Christiansen, K., & Knussmann, R. (1987). Sex hormones and cognitive functioning in men. *Neuropsychobiology, 18*, 27–36.

Churchland, P. M. (1984). *Matter and consciousness A contemporary introduction to the philosopy of mind*. Cambridge, MA: MIT Press.

Churchland, P. (1986). *Neurophilosophy: Towards a unified science of the mind-brain*. Cambridge, MA: MIT Press.

Corley, R., DeFries, J., Kuse, A., & Vandenberg, S. (1980). Familial resemblance for the identical blocks test of spatial ability: No evidence for X linkage. *Behavior Genetics, 10,* 211–216.

Couwenbergs, C., Knussmann, R., & Christiansen, K. (1986). Comparisons of the intra- and inter-individual variability in sex hormone levels of men. *Annals of Human Biology, 13,* 63–72.

Crockett, L., & Petersen, A. (1985). Pubertal status and psychosocial development: Findings from the early adolescence study. In R. Lerner & T. Foch (Eds.), *Biological–psychosocial interactions in early adolescence: A life span perspective.* Hillsdale, NJ: Erlbaum.

Cronbach, L. (1975). Beyond the two disciplines of scientific psychology. *American Psychologist, 30,* 116–127.

Dabbs, J., de La Rue, D., & Williams, P. (1991). Salivary testosterone and occupational choice: Actors, ministers, and other men. *Journal of Personality and Social Psychology, 59*(6), 1261–1265.

Dabbs, J., & Morris, R. (1990). Testosterone, social class, and antisocial behavior in a sample of 4,462 men. *Psychological Science, 1,* 209–211.

Dalton, K. (1976). Prenatal progesterone and educational attainments. *British Journal of Psychiatry, 129,* 438–442.

Dalton, K. (1979). Intelligence and prenatal progesterone: A reappraisal. *Journal of the Royal Society of Medicine, 72,* 97–399.

Danmarks Statistik (1985). *Kvinder & Maend* (Women & Men). København: Danmarks Statistik.

Denenberg, V., Berrebi, A., & Roslyn, H. (1988). Sex, brain, and learning differences in rats. *Behavioral and Brain Sciences, 11,* 188–189.

DeVoogd, T., & Nottebohm, F. (1981). Gonadal hormones induce dendritic growth in the adult avian brain. *Science, 214,* 202–204.

Diamond, M. (1988). *Enriching Heredity.* New York: Free Press.

Diamond, M., Dowling, G., & Johnson, R. E. (1981). Morphological cerebral cortical asymmetry in male and female rats. *Experimental Neurology, 71,* 261–268.

Döhler, K., & Hancke, J. (1978). Thoughts on the mechanism of sexual brain differentiation. In G. Dörner & M. Kawakami (Eds.), *Hormones and brain development.* Amsterdam: Elsevier/North-Holland Biomedical Press.

Döhler, K., Hancke, J., Srivastava, S., Hofmann, C., Shryne, J., & Gorski, P. (1984). Participation of estrogens in female sexual differentiation of the brain: Neuroanatomical, neuroendocrine and behavioral evidence. In G. Vries, J. Bruin, H. Uylings, & M. Corner (Eds.), *Sex differences in the brain: Progress in the brain research* (pp. 99–117). Amsterdam: Elsevier Biomedical Press.

Dor-Shav, N. (1976). In search of premenstrual tension: Note on sex-differences in psychological-differentiation as a function of cyclical physiological changes. *Perceptual and Motor Skills, 42,* 1139–1142.

Editorial, (1976). More divorce and less salary for women science PhDs. *New Scientist, 69,* 130–131.

Ehrhardt, A., Meyer-Bahlburg, H., Rosen, L., Feldman, J., Veridiano, N., & Zimmerman, I. (1985). Sexual orientation after prenatal exposure to exogenous estrogen. *Archives of Sexual Behavior, 14,* 57–77.

Ellis, L., & Ames, M. (1987). Neurohormonal functioning and sexual orientation: A theory of homosexuality–heterosexuality. *Psychological Bulletin, 101,* 233–258.

Ellis, L., & Nyborg, H. (1992). Racial/ethnic variations in male testosterone levels: A probable contributor to group differences in health. *Steroids, 57,* 72–75.

Ellis, L., Peckham, W., Ames, M., & Burke, D. (1988). Sexual orientation of human offspring may be altered by severe maternal stress during pregnancy. *The Journal of Sex Research, 25,* 52–157.

Eysenck, H. (1976). Ideology run wild. *American Psychologist, 31*(4), 311–312.

Farthing, M., Mattel, A., Edwards, C., & Dawson, A. (1982). Relationship between plasma testosterone and dihydrotestosterone concentrations and male facial hair growth. *British Journal of Dermatology, 107,* 559–564.

Feder, H. (1984). Hormones and sexual behavior. *Annual Review of Psychology, 35,* 165–200.

Garai, J., & Scheinfeld, A. (1968). Sex differences in mental and behavioral traits. *Genetic Psychology Monographs, 77,* 169–299.

Geschwind, N., & Galaburda, A. (1987). *Cerebral lateralization: Biological mechanisms, associations, and pathology.* Cambridge, MA: MIT Press.

Goldberg, S. (1977). *The inevitability of patriarchy.* London: Temple Smith.

Gooren, L. (1988). An appraisal of endocrine theories of homosexuality and gender dysphoria. In J. Sitsen (Ed.), *Handbook of sexology. Vol. 6: The pharmacology and endocrinology of sexual function.* Amsterdam: Elsevier Science Publishing.

Gordon, H., Corbin, E., & Lee, P. (1986). Changes in specialized cognitive function following changes in hormone levels. *Cortex, 22,* 399–415.

Gorski, R. (1974). The neuroendocrine regulation of sexual behavior. In G. Newton & A. Riesen (Eds.), *Advances in Psychobiology* (pp. 1–58). New York: John Wiley & Sons.

Gorski, R. (1984). Critical role for the medial preoptic area in the sexual differentiation of the brain. In G. Vries, J. Bruin, H. Uylings, & M. Corner (Eds.), *Sex differences in the brain: Progress in the brain research* (pp. 129–146). Amsterdam: Elsevier Biomedical Press.

Gouchie, C., & Kimura, D. (1990). The relation between testosterone levels and cognitive ability patterns. *Psychoneuroendocrinology, 16,* 1–30.

Gould, E., Woolley, C. S., & McEwen, B. S. (1991). Naturally occurring cell death in the developing dentate gyrus of the rat. *The Journal of Comparative Neurology, 304,* 408–418.

Goy, R., McEwen, B., Baker, S., Beatty, W., Czaja, J., Dörner, G., Ehrhardt, A., & Fox, T. (1980). *Sexual differentiation of the brain. Based on a work session of the neurosciences research program.* Cambridge: MA: MIT Press.

Greenwald, I., & Martinez-Arias, A. (1984). Programmed cell death in invertebrates. *Trends in NeuroSciences, 7,* 179–181.

Hampson, E. (1986). Variations in perceptual and motor performance related to phase of the menstrual cycle. *Canadian Psychology, 27,* 268.

Hampson, E. (1988). Variations in sex-related cognitive abilities across the menstrual cycle. Research Bulletin, 669. Department of Psychology, University of Western Ontario, Canada.

Hampson, E. (1990). Estrogen-related variations in human spatial and articulatory-motor skills. *Psychoneuroendocrinology, 15,* 97–111.

Hampson, E., & Kimura, D. (1988). Reciprocal effects of hormonal fluctuations on human motor and perceptuo-spatial skills. *Behavioral Neuroscience, 102,* 456–459.

Hapgood, F. (1979). *Why males exist: An inquiry into the evolution of sex.* New York: William Morrow and Company.

Harrell, T., & Harrell, M. (1945). Army General Classification Test scores for civilian occupations. *Educational and Psychological Measurement, 5,* 229–239.

Harris, L. (1979). Sex-related differences in spatial ability: A developmental psychological view. In C. Kopp (Ed.), *Becoming female, perspectives on development* (pp. 133–181). New York: Plenum Press.

Harris, L. (1980). Lateralized sex differences: Substrates and significance. *The Behavioral and Brain Sciences, 3,* 236–237.

Hinchliffe, J. (1981). Cell death in embryogenesis. In I. Bowen & R. Lockshin (Eds.), *Cell death in biology and pathology.* London: Chapman and Hall.

Hingley, P., & Cooper, C. (1983). The loners at the top. *New Society, 65,* 467–470.

Hofman, M. A., & Swaab, D. F. (1989). The sexually dimorphic nucleus of the preoptic area in the human brain: A comparative morphometric study. *Journal of Anatomy, 164,* 55–72.

Ho, H., Gilger, J., & Brink, T. (1986). Effects of menstrual cycle on spatial information-processes. *Perceptual and Motor Skills, 63,* 743–751.

Ho, K. C., Roessmann, U., Straumfjord, J. V., & Monroe, G. (1980). Analysis of brain weight: I. Adult brain weight in relation to sex, race, and age. *Archive for Pathology and Laboratory Medicine, 104,* 635–639.

Hoyenga, K., & Hoyenga, K. (1979). *The question of sex differences: Psychological, cultural, and biological issues.* Boston: Little, Brown and Company.

Hughes, R. (1983). Menstrual cycle influences on perceptual disembedding ability. *Perceptual and Motor Skills, 57,* 107–110.

Hyde, J., & Linn, M. (1986). *The psychology of gender.* Baltimore: Johns Hopkins University Press.

Jacklin, C. N. (1979). Epilogue. In M. A. Wittig & A. C. Petersen (Eds.), *Sex-related differences in cognitive functioning: Developmental issues* (pp. 357–371). New York: Academic Press.

Jacklin, C. (1989). Female and male: issues of gender. *American Psychologist, 44,* 127–133.

Jacob, F. (1982a). *The logic of life: A history of heredity.* New York: Pantheon Books.

Jacob, F. (1982b). *The possible and the actual.* New York: Pantheon Books.

Jacobson, C., Davis, F., & Gorski, R. (1985). Formation of the sexually dimorphic nucleus of the preoptic area: Neuronal growth, migration and changes in cell number. *Developmental Brain Research, 21,* 7–18.

Jardine, R., & Martin. N. (1983). Throwing accuracy and the genetics of spatial ability. *Behavior Genetics, 13,* 331–340.

Jensen, A. (1980). *Bias in mental testing.* New York: Free Press.

Jensen, A., & Reynolds, C. (1983). Sex difference on the WISC-R. *Personality and Individual Differences, 4,* 223–226.

Johnson, F. W. (1991). Biological factors and psychometric intelligence: A review. *Genetic, Social, and General Psychology Monographs, 117,* 3, 313–357.

Jonge, F. D., Louwerse, A., Ooms, M., Evers, P., Endert, E., & Van de Poll, N. (1989). Lesions of the SDN-POA inhibit sexual behavior of male Wistar rats. *Brain Research Bulletin, 23,* 483–492.

Juraska, J. (1984). Sex differences in developmental plasticity in the visual cortex and hippocampal dentate gyrus. In G. Vries, J. Bruin, H. Uylings, & M. Corner (Eds.), *Sex differences in the brain: Progress in the brain research* (pp. 205–214). Amsterdam: Elsevier Biomedical Press.

Juraska, J. (1986). Sex differences in developmental plasticity of behavior and the brain. In W. Greenough & J. Juraska (Eds.), *Developmental neuropsychobiology* (pp. 409–422). New York: Academic Press.

Juraska, J., Fitch, J., Henderson, C., & Rivers, N. (1985). Sex differences in the dendritic branching of dentate granule cells following differential experience. *Brain Research, 333,* 73–80.

Jurgens, H. W., Aune, I. A., & Pieper, U. (1990). *International data on anthropometry.* Geneva, Switzerland: International Labour Office.

Kaufman, A. (1975). Factor analysis of the WISC-R at 11 age levels between 6 and 16 years. *Journal of Consulting and Clinical Psychology, 43,* 135–147.

Kimura, D. (1983). Sex differences in cerebral organization for speech and praxic functions. *Canadian Journal of Psychology, 37,* 19–35.

Kimura, D. (1987). Are men's and women's brains really different? *Canadian Psychology, 28,* 133–147.

Kimura, D. (1991). Sex differences in cognitive function vary with the season. Research Bulletin #697. Department of Psychology, University of Western Ontario, Canada.

Kimura, D., & Harshman, R. (1984). Sex differences in brain organization for verbal and non-verbal functions. In G. Vries, J. Bruin, H. Uylings, & M. Corner (Eds.), *Sex differences in the brain: Progress in the brain research* (pp. 423–441). Amsterdam: Elsevier Biomedical Press.

Klaiber, E., Broverman, D., Vogel, W., & Kobayashi, Y. (1974). Rhythms in plasma MAO activity, EEG, and behavior during the menstrual cycle. In M. Ferin, F. Halberg, R. Richart, & R. Wiele (Eds.), *Biorhythms and human reproduction* (pp. 353–367). New York: John Wiley.

Knussmann, R., & Christiansen, K. (1986). Relations between sex hormone levels and sexual behavior in men. *Archives of Sexual Behavior, 15,* 429–445.

Knussmann, R., & Sperwien, A. (1988). Relations between anthropometric characteristics and androgen hormone levels in healthy young men. *Annals of Human Biology, 15,* 131–142.

Kolakowski, D., & Malina, R. (1974). Spatial ability, throwing accuracy and man's hunting heritage. *Nature, 251,* 410–412.

Kolb, B., Sutherland, R., Nonneman, A., & Whishaw, I. (1982). Asymmetry in the cerebral hemispheres of the rat, mouse, rabbit, and cat: The right hemisphere is larger. *Experimental Neurology, 78,* 348–359.

Komnenich, P., Land, D., Dickey, R., & Stone, S. (1978). Gonadal hormones and cognitive performance. *Physiological Psychology, 6,* 115–120.

Lacoste-Utamsing, C., & Holloway, R. (1982). Sexual dimorphism in the human corpus callosum. *Science, 216,* 1431–1432.

Landman, A., Sanford, L., Howland, B., Dawes, C., & Pritchard, E. (1976). Testosterone in human saliva. *Experienta, 32,* 940–941.

Lenney, E. (1979). Androgyny: Some audacious assertions toward its coming of age. *Sex Roles, 5,* 703–719.

Linn, M., & Petersen, A. (1985). Emergence and characterization of sex differences in spatial ability: A meta-analysis. *Child Development, 56,* 1479–1498.

Lynn, R. (1990). Negative correlations between verbal and visuo-spatial abilities: Statistical artifact or empirical relationship? *Personality and Individual Differences, 11,* 755–756.

Lynn, R., & Mulhern, G. (1991). A comparison of sex differences on the Scottish and American standardisation samples on the WISC-R. *Personality and Individual Differences, 12(11),* 1179–1182.

Maccoby, E., & Jacklin, C. (1978). *The psychology of sex differences Volume I: Text Volume II: Annotated bibliography.* Stanford, California: Stanford University Press.

Marshall, W., & Tanner, J. (1986). Puberty. In F. Falkner & J. Tanner (Eds.), *Human growth* (pp. 171–209). New York: Plenum Publishing Corporation.

Matarazzo, J. D., Bornstein, R. A., McDermott, P. A., & Noonan, J. V. (1986). Verbal IQ vs. performance IQ difference scores in males and females from the WAIS-R standardization sample. *Journal of Clinical Psychology, 42,* 965–974.

Matsumoto, S., Sato, I., Ito, T., & Matsuoka. A. (1966). Electroencephalographic changes during long term treatment with oral contraceptives. *International Journal of Fertility, 11,* 195–204.

McCall, R. (1977). Childhood IQs as predictors of adult, educational and occupational status. *Science, 197,* 482–483.

McCormick, C. M., & Witelson, S. F. (1991). A cognitive profile of homosexual men compared to heterosexual men and women. *Psychoneuroendocrinology, 6,* 459–473.

McEwen, B. (1987a). Observations on brain sexual differentiation: A biochemist's view. In J. Reinisch, L. Rosenblum, & S. Sanders (Eds.), *Masculinity/Femininity: Basic perspectives* (pp. 68–79). Oxford: Oxford University Press.

McEwen, B. (1987b). Steroid hormones and brain development: Some guidelines for understanding actions of pseudohormones and other toxic agents. *Environmental Health Perspectives, 74,* 177–184.

McEwen, B. (1988a). Actions of sex hormones on the brain: "Organization" and "activation" in relation to functional teratology. In G. Boer, M. Feenstra, M. Mirmiram, D. Swaab, & F. V. Haaren (Eds.), *Progress in brain research* (pp. 121–134). Amsterdam: Elsevier.

McEwen, B. (1988c). Genomic regulation of sexual behavior. *Journal of Steroid Biochemistry, 30,* 179–183.

McEwen, B. (1988d). Steroid hormones and the brain: Linking "nature" with "nurture." *Neurochemical Research, 13,* 663–669.

McEwen, B. (1991a). Our changing ideas about steroid effects on an ever-changing brain. *Seminars in the Neurosciences, 3,* 497–507.

McEwen, B. (1991b). Steroids affect neural activity by acting on the membrane and the genome. *Trends in Pharmacological Sciences, 12,* 141–147.

McEwen, B., Jones, K., & Pfaff, D. (1987). Hormonal control of sexual behavior in the female rat: Molecular, cellular and neurochemical studies. *Biology of Reproduction, 36,* 37–45.

McGee, M. (1979). Human spatial abilities: Psychometric studies and environmental, genetic, hormonal, and neurological influences. *Psychological Bulletin, 86,* 889–917.

McGlone, J. (1986). The neuropsychology of sex differences in human brain organization. In G. Goldstein & R. Tarter (Eds.), *Advances in Clinical Neuropsychology* (pp. 1–29). New York: Plenum Press.

McGlone, J. (1980). Sex differences in human brain asymmetry: A critical survey. *The Behavioral and Brain Sciences, 3*, 215–263.

McKeever, W. (1986). The influences of handedness, sex, familial sinistrality and androgyny on language laterality, verbal ability, and spatial ability. *Cortex, 22*, 521–537.

Meikle, A., Bishop, D., Stringham, J. D., & West, D. (1986). Quantitating genetic and nongenetic factors that determine plasma sex-steroid variation in normal male twins. *Metabolism, 35*, 1090–1095.

Money, J., Schwartz, M., & Lewis, V. (1984). Adult erotosexual status and fetal hormonal masculinization and demasculinization: 46,XX congenital virilizing adrenal hyperplasia and 46,XY androgen-insensitivity syndrome compared. *Psychoneuroendocrinology, 9*, 405–414.

Money, J. (1987). Sin, sickness, or status? Homosexual gender identity and psychoneuroendocrinology. *American Psychologist, 42*, 384–389.

Monod, J. (1971). *Chance & Necessity.* New York: Knopf.

Moss, R., & Dudley, C. (1984). Molecular aspects of the interaction between estrogen and the membrane excitability of hypothalamic nerve cells. In G. Vries, J. Bruin, H. Uylings, & M. Corner (Eds.), *Sex differences in the brain: Progress in brain research* (pp. 3–22). Amsterdam: Elsevier Biomedical Press.

Nass, R., Baker, S., Speiser, P., Virdis, R., Balsamo, A., Cacciari, E., Loche, A., Dumic, M., & New, M. (1987). Hormones and handedness: Left-hand bias in female congenital adrenal hyperplasia patients. *Neurology, 37*, 711–715.

Nieschlag, E. (1974). Circadian rhythm of plasma testosterone. In J. Aschoff, F. Ceresa, & F. Halberg (Eds.), *Chronobiological aspects of endocrinology* (pp. 117–128). New York: Schattaur Verlag.

Norman, A. W., & Litwack, G. (1987). *Hormones.* San Diego: Academic Press.

Nottebohm, F. (1980). A continuum of sexes bedevils the search for sexual differences? *The Behavioral and Brain Sciences, 3*, 245–246.

Nottebohm, F. (1981). A brain for all seasons: Cyclical anatomical changes in song control nuclei of the canary brain. *Science, 214*, 1368–1370.

Nottebohm, F. (1989). From bird song to neurogenesis. *Scientific American, 260*(2), 56–61.

Nottebohm, F., Nottebohm, M., & Crane, L. (1986). Developmental and seasonal changes in canary song and their relation to changes in the anatomy of song-control nuclei. *Behavioral and Neural Biology, 46*, 445–471.

Nyborg, H. (1977). *The rod-and-frame test and the field dependence dimension: Some methodological, conceptual, and developmental considerations.* Copenhagen: Dansk Psykologisk Forlag.

Nyborg, H. (1979, June). *Sex chromosome abnormalities and cognitive performance V: Female sex hormone and discontinuous cognitive development.* Paper and hand-out presented in the Symposium on Cognitive Studies at the Fifth Biennial Meeting of the International Society for the Study of Behavioral Development, Lund, Sweden.

Nyborg, H. (1983). Spatial ability in men and women: Review and new theory. *Advances in Human Research and Therapy, 5*(whole issue), 39–140.

Nyborg, H. (1984). Performance and intelligence in hormonally-different groups. In G. D. Vries, J. D. Bruin, H. Uylings, & M. Corner (Eds.), *Sex differences in the brain. Progress in brain research* (pp. 491–508). Amsterdam: Elsevier Biomedical Press.

Nyborg, H. (1987a). Individual differences or different individuals? That is the question. *Behavioral and Brain Sciences, 10*, 34–35.

Nyborg, H. (1987b, June). *Covariant trait development across races and within individuals: Differential K theory, genes, and hormones.* Paper presented in the Symposium on Biology-Genetics at the Third Meeting of the International Society for the Study of Individual Differences, Toronto, Canada.

Nyborg, H. (1988a). Change at puberty in spatioperceptual strategy on the rod-and-frame test. *Perceptual and Motor Skills, 67*, 129–130.

Nyborg, H. (1988b). Sex hormones and covariant body, brain and behavioral development. *Neuroendocrinology Letters (Abstracts), 10*, 217.

Nyborg, H. (1989). The nature of Nature–Nurture interaction. *Behavior Genetics, 20*, 738–739.

Nyborg, H. (1990a). Sex hormones, brain development, and spatio-perceptual strategies in Turner's syndrome. In D. Berch & B. Bender (Eds.), *Sex chromosome abnormalities and human behavior: Psychological studies* (pp. 100–128). Boulder, CO: Westview Press.

Nyborg, H. (1990b). Good, bad, and ugly questions about heredity. *Behavioral and Brain Sciences, 13*, 142–143.

Nyborg, H. (1994a). Individual, sex, and race differences in body, brain, and specific abilities: The General Trait Covariance–Androgen/Estrogen model for development. (Submitted).

Nyborg, H. (1994b). *Hormones, sex, and society: The physiological approach.* New York: Greenwood Publishing Co. (in press).

Nyborg, H. (1994c). Gonadal hormones, body, brain, intellectual, and personality development at puberty: A 14-year cohort-sequential study of 8–18-year-old, unselected Danish children. Unpublished manuscript.

Nyborg, H. (1992, April). *Genes, hormones, maturation, and the etiology of psychoses: An application of the GTC–A/E model.* Paper presented at the International Symposium on Rate of Maturation, Brain Development, and Behaviour," University of Trondheim, Geilo, Norway.

Nyborg, H., & Bøggild, C. (1989). Mating behavior: Moves of mind or molecules? *The Behavioral and Brain Sciences, 12*, 29–30.

Nyborg, H., & Nielsen, J. (1977). Sex chromosome abnormalities and cognitive performance. III: Field dependence, frame dependence, and failing development of perceptual stability in girls with Turner's syndrome. *Journal of Personality, 96*, 205–211.

Nyborg, H., & Nielsen, J. (1981a). Spatial ability of men with karyotype 47,XXY, 47,XYY, or normal controls. In W. Schmid & J. Nielsen (Eds.), *Human behavior and genetics* (pp. 85–106). Amsterdam: Elsevier/North-Holland Biomedical Press.

Nyborg, H., & Nielsen, J. (1981b). Sex hormone treatment and spatial ability in women with Turner's syndrome. In W. Schmid & J. Nielsen (Eds.), *Human behavior and genetics* (pp. 167–182). Amsterdam: Elsevier/North-Holland Biomedical Press.

Nyborg, H., & Sommerlund, B. (1992). Spearman's *g*, the verbal-performance balance, and brain processes: The Lynn-Vernon debate. *Personality and Individual Differences, 13*(11), 1253–1255.

O'Connor, J. (1943). *Structural Visualization.* Boston: Human Engineering Lab.

O'Kusky, J., Strauss, E., Kosaka, B., Wada, J., Li, D., Druhan, M., & Petrie, J. (1988). The corpus callosum is larger with right-hemisphere cerebral speech dominance. *Annals of Neurology, 24*, 379–383.

Pavlides, C., Westlind-Danielsson, A., Nyborg, H., & McEwen, B. (1991). Neonatal hyperthyroidism disrupts hippocampal LTP and spatial learning. *Experimental Brain Research, 85*, 559–564.

Petersen, A. (1976). Physical androgyny and cognitive functioning in adolescence. *Developmental Psychology, 12*, 524–533.

Petersen, A., & Crockett, L. (1985). Pubertal timing and grade effects on adjustment. Special Issue: Time of maturation and psycho-social functioning in adolescence: I. *Journal of Youth and Adolescence, 14*(3), 191–206.

Petersen, A., & Taylor, B. (1980). The biological approach to adolescence: Biological change and psychological adaptation. In J. Adelson (Ed.), *Handbook of Adolescent Psychology* (pp. 117–155). New York: John Wiley & Sons.

Plomin, R., & Daniels, D. (1987). Why are children in the same family so different from one another? *Behavioral and Brain Sciences, 10*, 1–60.

Purifoy, F. (1981). Endocrine-environment interaction in human variability. *Annual Review of Anthropology, 10*, 141–162.

Purifoy, F., & Koopmans, L. (1979). Androstenedione, testosterone, and free testosterone concentration in women of various occupations. *Social Biology, 26,* 179–188.

Ragins, B., & Sundstrom, E. (1989). Gender and power in organizations: A longitudinal perspective. *Psychological Bulletin, 105,* 51–88.

Read, G., Riad-Fahmy, D., Walker, R. F., & Griffiths, K. (Eds.) (1984). Immunoassays of steroids in saliva. *Proceedings of the Ninth Tenovus Workshop,* Cardiff, Wales: Alpha Omega Publ.

Reid, I. (1982). Vital statistics. In I. Reid & E. Wormald (Eds.), *Sex differences in Britain* (pp. 29–58). London: Grant McIntyre.

Reinberg, A., & Lagoguey, M. (1978). Circadian and circannual rhythms in sexual activity and plasma hormones (FSH, LH, testosterone) of five human males. *Archives for Sexual Behavior, 7,* 13–30.

Reinberg, A., Lagoguey, M., Chauffournier, J., & Cesselin, F. (1975). Circannual and circadian rhythms in plasma testosterone in 5 healthy young Parisian males. *Acta Endocrinologica, 80,* 732–743.

Romano, G. J., Mobbs, C. V., Lauber, A., Howells, R. D., & Pfaff, D. W. (1990). Differential regulation of proenkephalin gene expression by estrogen in the ventromedial hypothalamus of male and female rats: Implications for the molecular basis of a sexually differentiated behavior. *Brain Research, 536,* 63–68.

Ross, R., Bernstein, L., Judd, L., Hanisch, R., Pike, M., & Henderson, B. (1986). 1986 serum testosterone levels in healthy young black and white men. *Journal of the National Cancer Institute, 76,* 45–48.

Roth, J., LeRoith, D., Shiloach, J., Rosenzweig, J., Leisniak, M., & Havrankova, J. (1982). The evolutionary origins of hormones, neurotransmitters and other extracellular chemical messengers. *The New England Journal of Medicine, 306,* 523–527.

Runyan, W. (1983). Idiographic goals and methods in the study of lives. *Journal of Personality, 51,* 413–437.

Rushton, J. P. (1992). Cranial capacity related to sex, rank, and race in a stratified random sample of 6,325 military personnel. *Intelligence, 16,* 401–413.

Rushton, J. P. (1992b). Evidence for race and sex differences in cranial capacity from International labor office data. Research Bulletin, 712. Department of Psychology, University of Western Ontario, Canada.

Sandhu, S., Cook, P., & Diamond, M. (1986). Rat cerebral cortical estrogen receptors: Male-female, right-left. *Experimental Neurology, 92,* 186–196.

Sannikka, E., Terho, P., Suominen, J., & Santti, R. (1983). Testosterone concentrations in human seminal plasma and saliva and its correlation with non-protein-bound and total testosterone levels in serum. *International Journal of Andrology, 6,* 319–330.

Schapiro, S., Salas, M., & Vukovich, K. (1970). Hormonal effects on ontogeny of swimming ability in the rat: Assessment of central nervous system development. *Science, 168,* 147–150.

Schindler, G. L. (1979). Testosterone concentration, personality patterns, and occupational choice in women. *Dissertation Abstracts International, 40,* 1411A, (University Microfilms No. 79-19,403).

Sherman, J. (1967). Problem of sex differences in space perception and aspects of intellectual functioning. *Psychological Review, 74,* 290–299.

Shute, V., Pellegrino, J., Hubert, L., & Reynolds, R. (1983). The relationship between androgen levels and human spatial abilities. *Bulletin of the Psychonomic Society, 21,* 465–468.

Silverman, I., & Phillips, K. (1991, August). *Effects of estrogen during the menstrual cycle on spatial performance.* Paper presented at the Meeting of the Human Behavior and Evolution Society, McMaster University, Hamilton, Ontatio, Canada.

Smals, A., Kloppenborg, P., & Benraad, T. (1976). Circannual cycle in plasma testosterone levels in man. *Journal of Clinical Endocrinology and Metabolism, 42,* 979–982.

Smith, I. (1964). *Spatial ability: Its educational and social significance.* San Diego, California: Robert R. Knapp.

Spearman, C. (1904). General intelligence, objectively determined and measured. *American Journal of Psychology, 15,* 201–293.

Spence, J., & Helmreich, R. (1979). On assessing "Androgyny." *Sex Roles, 5,* 721–737.

Stafford, R. (1961). Sex differences in spatial visualization as evidence of sex-linked inheritance. *Perceptual and Motor Skills, 13,* 428.

Stafford, R. (1963). An investigation of similarities in parent-child test scores for evidence of hereditary components. Educational Testing Service, Princeton.

Stafford, R. (1965, April). *Spatial visualization and quantitative reasoning: One gene or two?* Paper presented at Eastern Psychological Association, Atlantic City, NJ.

Steward, J., & Kolb, B. (1988). The effects of neonatal gonadectomy and prenatal stress on cortical thickness and asymmetry in rats. *Behavioral and Neural Biology, 49,* 344–360.

Swaab, D., & Fliers, E. (1985). A sexually dimorphic nucleus in the human brain. *Science, 228,* 1112–1115.

Swaab, D., & Hofman, M. (1988). Sexual differentiation of the human hypothalamus: Ontogeny of the sexual dimorphic nucleus of the preoptic area. *Developmental Brain Research, 44,* 314–318.

Swaab, D., Hofman, M., & Fisser, B. (1988). Sexual differentiation of the human brain. *Neuroendocrinology Letters (Abstracts), 10,* 4, 217.

Symond, D. (1979). *The evolution of human sexuality.* New York: Oxford University Press.

Tan, Ü., & Akgün, A. (1992). There is a direct relationship between nonverbal intelligence and serum testosterone level in young men. *International Journal of Neuroscience, 64,* 213–216.

Tanner, J. (1975). Growth and endocrinology of the endocrinology of the adolescent. In L. Gardner (Ed.), *Endocrine and genetic diseases of childhood and adolescence* (pp. 14–64). Toronto: W. B. Saunders.

Tarttelin, M., & Gorski, R. (1988). Postnatal influence of diethylstilbestrol on the differentiation of the sexually dimorphic nucleus in the rat is as effective as perinatal treatment. *Brain Research, 456,* 271–274.

Terman, L. (1925). Mental and physical traits of a thousand gifted children. In L. Terman (Ed.), *Genetic studies of genius.* Stanford, CA: Stanford University Press.

Toran-Allerand, C. (1984). On the genesis of sexual differentiation of the central nervous system: Morphogenetic consequences of steroidal exposure and possible role of alpha-fetoprotein. In G. Vries, J. Bruin, H. Uylings, & M. Corner (Eds.), *Sex differences in the brain: Progress in the brain research* (pp. 63–98). Amsterdam: Elsevier Biomedical Press.

Toran-Allerand, D. (1986). Sexual differentiation of the brain. In W. Greenough & J. Juraska (Eds.), *Developmental neuropsychobiology* (pp. 175–211). New York: Academic Press.

Truman, J. (1985). Hormonal approaches to the study of cell death in a developing nervous system. In G. Edelman, W. Gall, & W. Cowan (Eds.), *Molecular bases of neural development* (pp. 531–545). New York: John Wiley & Sons.

Truman, J., & Schwartz, L. (1984). Steroid regulation of neuronal death in the moth nervous system. *The Journal of Neuroscience, 4,* 274–280.

Van Valen, L. (1974). Brain size and intelligence in man. *American Journal of Physical Anthropology, 40,* 417–424.

Vandenberg, S., & Kuse, A. (1979). Spatial ability: A critical review of the sex-linked major gene hypothesis. In M. Wittig & A. Petersen (Eds.), *Sex-related differences in cognitive functioning: Developmental issues* (pp. 67–95). New York: Academic Press.

Vernon, P. A. (1990). The effect of holding *g* constant on the correlation between verbal and nonverbal abilities: A comment on Lynn's "The intelligence of the Mongoloids" (1987). *Personality & Individual Differences, 11,* 751–754.

Vining, D. J. (1982). On the possibility of the reemergence of a dysgenic trend with respect to intelligence in American fertility differentials. *Intelligence, 6,* 241–264.

Vining, D. (1984). Subfertility among the very intelligent: An examination of the American mensa. *Personality and Individual Differences, 5,* 725–733.

Vogel, S. (1990). Gender differences in intelligence, language, visuo-motor abilities and academic achievement in students with learning disabilities: a review of the literature. *Journal of Learning Disability, 23,* 44–52.

Vries, G. D., Bruin, J. D., Uylings, H., & Corner, M. E. (Eds.) (1984). *Sex differences in the brain: Relation between structure and function.* Amsterdam: Elsevier Science Publishers.

Waber, D. (1976). Sex differences in cognition: A function of maturation rate. *Science, 192,* 572–574.

Waber, D. (1977a). Sex differences in mental abilities, hemispheric lateralization, and rate of physical growth at adolescence. *Developmental Psychology, 13,* 29–38.

Waber, D. (1977b). Biological substrates of field dependence: Implications of the sex difference. *Psychological Bulletin, 84,* 1076–1087.

Waber, D. (1979). Cognitive abilities and sex-related variations in the maturation of cerebral cortical functions. In M. Wittig & A. Petersen (Eds.), *Sex-related differences in cognitive functioning* (pp. 161–186). New York: Academic Press.

Wada, J., Clarke, R., & Hamm. A. (1975). Cerebral hemispheric asymmetry in humans. *Archives of Neurology, 32,* 239–246.

Wahlsten, D. (1990). Insensitivity of the analysis of variance to heredity–environment interaction. *Behavioral and Brain Sciences, 13,* 109–161.

Waller, J. (1971). Achievement and social mobility: Relationships among IQ score, education, and occupation in two generations. *Social Biology, 18,* 252–259.

Watson, N., & Kimura, D. (1990). *Nontrivial sex differences in throwing and intercepting: Relation to psychometrically defined spatial functions.* (Research Bulletin #693). London, ON: Department of Psychology, University of Western Ontario.

Wechsler, D. (1981). *WAIS-R Manual: Wechsler Adult Intelligence Scale-Revised.* New York: Psychological Corporation.

Whitam, F., & Mathy, R. (1988). Childhood cross-gender behavior of lesbians in four societies: Brazil, Peru, the Philippines, and the United States. *Archives of Sexual Behavior, 20*(2), 151–170.

Willerman, L., Schultz, R., Rutledge, J. N., & Bigler, E. D. (1991). In vivo brain size and intelligence. *Intelligence, 15,* 223–228.

Witelson, S., & Pallie, W. (1973). Left hemisphere specialization for language in the newborn: Neuroanatomical evidence of asymmetry. *Brain, 96,* 641–646.

Witkin, H. (1973). The role of cognitive style in academic performance and in teacher-student relations. *Educational Testing Service Research Bulletin RB-73-11.* Princeton, NJ.

Witkin, H., Dyk, R., Faterson, H., Goodenough, D., & Karp, S. (1962). *Psychological differentiation.* Potomac, MD: Erlbaum.

Witkin, H., & Goodenough, D. (1977). Field dependence and interpersonal behavior. *Psychological Bulletin, 84,* 661–689.

Witkin, H., Goodenough, D., & Karp, S. (1967). Stability of cognitive style from childhood to young adulthood. *Journal of Personality and Social Psychology, 7,* 291–300.

Witkin, H., Goodenough, D., & Oltman, P. (1977). Psychological differentiation: Current status. *Educational Testing Service Research Bulletin RB-77-17,* Princeton, NJ.

Witkin, H. A., Lewis, H. B., Hertzman, M., Machover, K., Meissner, P. B., & Wapner, S. (1954). *Personality through perception.* New York: Harper.

Witkin, H., Moore, C., Goodenough, D., & Cox, P. (1977). Field-dependent and field-independent cognitive styles and their educational implications. *Review of Educational Research, 47,* 1–64.

Witzmann, R. (1981). *Steroids: Keys to Life.* New York: Van Nostrand Reinhold.

Woodfield, R. (1984). Embedded figure test performance before and after childbirth. *British Journal of Psychology, 75,* 81–88.

Woodfield, R., Whitehead, M., & Asso, D. (1987). Performance on particular tests of spatial functioning during the early follicular and peri-ovulatory periods of the menstrual cycle. *Gynecological Endocrinology (Abstracts), 1,* 85.

Woolley, C., Gould, E., Frankfurt, M., & McEwen, B. (1990). Naturally occuring fluctuations in dendritic spine density on adult hippocampal pyramidal neurons. *The Journal of Neuroscience, 10,* 4035–4039.

Wormald, E. (1982). Political participation. In I. Reid & E. Wormald (Eds.), *Sex differences in Britain* (pp. 175–203). London: Grant McIntyre.

Young, M., & Reeve, T. (1980). Discriminant analysis of personality and body-image factors of females differing in percent body fat. *Perceptual and Motor Skills, 50,* 547–552.

Ziegler, N. I., & Lichtensteiger, W. (1992). Asymmetry of brain aromatase activity: Region- and sex-specific developmental patterns. *Neuroendocrinology, 55,* 512–518.

Section II
Information Processing and Memory

5

Neuropsychological Models of Information Processing
A Framework for Evaluation

W. Grant Willis and Andrew D. Aspel

Many different neuropsychological models of information processing have been proposed. The research that has provided the empirical basis for these models is diverse; it addresses a variety of questions and it approaches the neurological basis of information processing from several levels of analysis.

Hiscock and Kinsbourne (1987) classified three such levels of analysis as anatomical, physiological, and behavioral. The constructs that are germane to each level include neuroanatomical structures, brain functions, and perception and cognition, respectively. Clearly, neuropsychological models of information processing require synthesis across all levels of analysis, because all associated constructs are important in their development. Thus, this classification is especially useful in these instances because it emphasizes the degree of inference used in generalizing research and clarifies the kinds of interpretations that are appropriate from particular research procedures and paradigms. For example, research associated with a given level of analysis sometimes inappropriately is generalized to other levels of analysis, such as when neuroanatomical substrata are inferred from behavioral data. These kinds of errors can lead to inaccurate conclusions and ill-conceived models.

Particular research procedures align with each level of analysis. At the anatomical level, for example, procedures such as computed tomography (CT) and autopsy are informative. At the physiological level, procedures such as positron emission tomography (PET), regional cerebral blood flow, and evoked cortical responses are informative (Hynd & Willis, 1988). Finally, at the behavioral level, noninvasive procedures are informative, such as dichotic listening, in which different auditory stimuli simultaneously are presented to each ear; tachistoscopic procedures, in which visual stimuli are presented to the visual hemifields more quickly than eye movements can occur; and dichaptic procedures, in which objects are manipulated for tactual identification (Hannay, 1986; Jeeves & Baumgartner, 1986). All of these procedures yield measurable response variables that subsequently can be interpreted with varying levels of inference.

The purpose of this chapter is to provide a framework from which to evaluate contemporary neuropsychological models of cognitive processing. Selected models that have well-established empirical support at each level of analysis are introduced as illustrations. These models are organized in terms of their potential complexity as assessed by the number of dimensions upon which they are based. For example, Hughlings Jackson's three Cartesian coordinates that traditionally have been used for mapping the human central nervous system are (1) the x-axis, which corresponds to the saggital plane and demarcates left from right divisions, (2) the y-axis, which corresponds to the horizontal plane and demarcates levels of neuraxis, and (3) the z-axis, which corresponds to the coronal or facial plane and demarcates anterior from posterior divisions. Some neuropsychological models of information processing exclusively focus on only one of these dimensions, others focus on two, and still others on all three. Examples of each are introduced, implications for individual differences are suggested, and differences and similarities among these models are identified.

I. SELECTED MODELS

A. Unidimensional Models

The unidimensional models primarily have focused on the saggital plane, emphasizing the relationship between the left and the right cerebral hemispheres. Clearly, the hemispheres do not process all kinds of information with equal efficiency. It was once assumed that only higher order cognitive processes were subject to hemispheric specialization (e.g., Luria, 1973; Moscovitch, 1979), but it now appears that even some sensory processes may be lateralized to a particular hemisphere, albeit less strongly (Davidoff, 1982; Kimura & Durnford, 1974). Thus, the relative superiority of one hemisphere over the other in terms of processing specific cognitive tasks has become a popular topic for speculation even among the lay public. A common misconception, however, is that because the cerebral hemispheres are specialized differentially for particular functions, information processing mediated at the cortical level of neuraxis is disjointed rather than cooperative.

Most models that address the neuropsychological basis of information processing as it relates to cerebral specialization have been based on perceptual asymmetries that have been measured through procedures conducted at the behavioral level of analysis. Essentially, they can be classified as either afferent or efferent models (also see Moscovitch, 1986). Afferent models emphasize the largely contralateral sensorineural projection pathways that link a particular receptor organ (e.g., retina of the eye) with the cortical region specialized for the corresponding modality (e.g., visual cortex in the occipital lobe). In this sense, it is instructive to consider afferent models as bottom-up (or data-based) models. In contrast, efferent models emphasize the role of the cerebral cortex for directing attention and for guiding the perception of environmental stimuli. Thus, these models often are considered as top-down (or conceptually based) models because they suggest a more active role in the perceptual process.

A primary goal of all afferent and efferent models is to synthesize principles of neurology with those of cognitive psychology. Afferent models traditionally have been more closely aligned with the neurosciences, whereas efferent models traditionally have been more closely aligned with the cognitive sciences. As Kandel (1985) so aptly articulated, however, the boundary between the disciplines of biology and behavior are arbitrary and changing. Thus, it now appears that our potential for a better understanding of the neurological basis of cognition will lie in the interrelationships between these two broad disciplines.

1. Afferent Models   One assumption of afferent models is that the processing of information comprises a number of hierarchically arranged components, or operations, each of which receives its input from the previous operation. The execution of particular operations, at least at some stages, further is assumed to occur within the cerebral hemispheres. According to afferent models, there are two possible explanations for the asymmetries that are elicited from behavioral procedures, such as dichotic listening and dichaptic and tachistoscopic tests.

One explanation suggests that each cerebral hemisphere is capable of executing the operations required, but that the hemispheres neither execute those operations identically nor with the same degree of efficiency. For example, if Task X comprises a series of Operations, a, b, and c, this explanation suggests that either hemisphere is capable of executing those operations. Any particular operation (e.g., Operation a), however, may be executed differently or more efficiently in one hemisphere (e.g., the left hemisphere) than in the other (e.g., the right hemisphere). Thus, if Task X were projected to the left hemisphere, which more efficiently executes Operation a, then the behavioral response to Task X would be more favorable than if Task X were projected to the right hemisphere. Thus, a perceptual asymmetry would result that favors the left cerebral hemisphere, that is, the contralateral (or right-sided) hemispace. Given the largely (and in the case of the visual system, the exclusively) contralateral sensorineural pathways, the functional specialization of a cerebral hemisphere for processing this information would be inferred from the perceptual asymmetry.

An alternative explanation suggests that at least one of the operations required for processing the information can only be executed in one particular hemisphere. Thus, if a cognitive task initially were projected to the cerebral hemisphere that could not execute a component operation, then the information would be transferred to the other hemisphere across commissural (i.e., interhemispheric) pathways. For example, assume that Operation *a* only can be executed in the left hemisphere. In this instance, if Task X initially were projected to the right hemisphere, then information would need to be transferred to the left hemisphere in order to execute Operation *a*. This explanation assumes that operations that require interhemispheric transfer are executed less rapidly or less accurately than operations that do not require interhemispheric transfer. In this example, a perceptual asymmetry would result that favors the left cerebral hemisphere. Again, given the contralateral sensorineural pathways, the functional specialization of a cerebral hemisphere for the information processing is inferred from the perceptual asymmetry. Thus, for both of these explanations of afferent models, it could be argued that the left cerebral hemisphere functionally was specialized for information processing associated with Task X.

Moscovitch (1986) delineated several predictions that follow from these two explanations of afferent models and described methods to distinguish between them. Clearly, there are implications of these divergent explanations for understanding the operations involved in information processing. Additionally, there are important clinical implications. Here, different psychoeducational treatment options might be selected depending on which explanation better accounts for a particular cognitive task.

2. Efferent Models   One of the earliest efferent models arose from Kinsbourne's (1970, 1973, 1975) selective activation hypothesis. This hypothesis, which subsequently was incorporated into a revised theory of functional cerebral distance (Kinsbourne & Hicks, 1978), suggested that perceptual asymmetries result from the differential activation of the cerebral hemispheres. This differential activation is due to the involvement of the hemispheres in a secondary task, rather than, for example, to differences in processing capacity. Because the sensorineural pathways are organized contralaterally, differential activation of the hemispheres biases attention in favor of the hemispace that is contralateral to the more activated cerebral hemisphere. Thus, stimuli within that hemispace are processed more efficiently than stimuli within the ipsilateral hemispace. Currently, the revised model and alternative theories probably better account for the accumulated relevant data, for example, of the pervasive interference rather than facilitation effects of secondary tasks (Kinsbourne & Hiscock, 1983a), but the selective activation hypothesis is recognized as an important forerunner to other efferent models of cerebral specialization.

One of these alternative theories was proposed by Friedman and Polson (1981). These investigators conceptualized each cerebral hemisphere as an independent, limited resource system. Resources refer to theoretical mechanisms responsible for the execution of various component operations of a task. In this model, the resources comprised by each cerebral hemisphere are finite. Thus, the differential executions of operations within a given cerebral hemisphere conceivably need to compete for this limited supply of resources.

This model assumes that each cerebral hemisphere has the same capacity limit, and that increased activation of one hemisphere always is accompanied by identically increased activation of the opposite hemisphere. Finally, although the hemispheres are conceptualized as separate systems, the products of those systems may become available to the opposite hemisphere as input via commissural transfer. This theory, of course, is in obvious contradistinction to the selective activation hypothesis that holds as its major premise that the hemispheres can be activated differentially.

Evidence to support this theoretical model has accrued primarily from studies in which two concurrently performed tasks are processed. When the component operations of one of those tasks, X, primarily are executed in one cerebral hemisphere, there is a relatively limited amount of resources available for the concurrent processing of the second task, Y, when the component operations associated with Y occur within the homologous hemisphere. Of course, this is not the case when the component operations associated with Y primarily occur within the opposite hemisphere. Thus, perceptual asymmetries associated with Y theoretically reflect hemispheric differences in processing efficiency. Functional lateralization of X, the primary task, therefore, is assumed.

Kinsbourne's (1982) theory of hemispheric specialization, that is, his principle of functional cerebral distance, provides an alternative efferent model of cerebral specialization in marked contrast to the limited-capacity conceptualization of Friedman and Polson (1981). From the perspective of Kinsbourne's theory, the assumption of a finite pool of resources is unnecessary. Instead, the brain is conceptualized as a differentiated neural network that comprises interconnected components responsible for particular cognitive operations. Given this conceptualization, it follows that information processing necessarily involves a larger portion of the total cerebral space than simply its locus of initiation. Consequently, concurrently processed tasks are predicted to conflict to the extent that their component operations overlap.

The principle of functional cerebral distance (Kinsbourne & Hicks, 1978) has received empirical support again primarily at the behavioral level of analysis from research specifically designed to test its associated hypotheses and reinterpretations of retrospectively viewed data. For example, Kinsbourne and Hicks (1978) reviewed studies of simultaneous imitative effects between contralateral limbs and between speech and manual motor behaviors. Subsequently, Kinsbourne and Hiscock (1983a) reviewed studies of competition between (1) two output processes, (2) an output and a cognitive process, (3) two input processes, and (4) an input and a cognitive process. In these instances, output processes included sequential finger movements, finger oscillation, and expressive speech. Cognitive processes included reading, memory encoding, and visual scanning. Input processes included stimuli contralaterally directed to a particular cerebral hemisphere through dichotic listening and tachistoscopic tests. Results of this research suggest that relative amounts of motor overflow, transfer, and interference can be explained in terms of degree of overlap among the component operations involved in the information processing.

Thus, Kinsbourne (1982) speculated that highly connected cerebral regions, which probably subserve similar processes, may lend themselves to successive (as opposed to simultaneous) information processing to permit the most efficient execution of any particular cognitive task. Conversely, regions of widely disparate functional cerebral distance, which probably subserve dissimilar processes, may be better suited for relatively more simultaneous information processing with respect to executing cognitive tasks. In this sense, cerebral specialization is viewed as a kind of neural separation between complementary component operations that, subsequent to sufficient elaboration, ultimately are aggregated to result in a unitary pattern of behavior. The primary advantage of this initial separation of component operations is that it protects the distinct, but complementary, contributions of various operations from mutual interference.

These and other efferent theoretical models have some similarities, but they also comprise important differences. The currently available data are insufficient to select one model as most appropriate for organizing the research on cerebral hemispheric specialization as it relates to information processing. Issues to be addressed include the so-called independence of cerebral hemispheres, the extent to which cerebral resources are differentiated, and, perhaps most important, the integrative functions of the cerebral hemispheres associated with higher order information processing. Some research, for instance, suggests that the particular hemisphere in which the component operations of a task are executed may be less important than the protection that the separate hemispheres provide against interference among those operations (Merola & Liederman, 1985). This is an intriguing hypothesis because it challenges the long-standing view of specific specializations of the cerebral hemispheres. Clearly, further empirical investigation is warranted.

B. Bidimensional Models

Perhaps one of the most influential neuropsychological models of information processing was introduced by Luria (1970). In contrast to the unidimensional models identified, Luria's bidimensional model relatively (although not completely) deemphasises the saggital plane, and instead focuses on the horizontal and coronal planes. This model also provides an excellent example of how research conducted at the anatomical and behavioral levels of analyses can be synthesized.

Luria (1970) conceptualized three basic functional units of the brain: (1) the subcortical, (2) the posterior cortical, and (3) the anterior cortical. Each unit is distinctive in terms of its differentiated function. There are extensive interconnections within and among the units, however, and these interconnections provide the anatomical basis for the mutual interdependence of the units. Each unit both influences and is influenced by the other units. Luria emphasized that all three units always are involved in the performance of any behavior, an assertion that clearly militates against a strictly localizationist perspective on the neuropsychology of information processing.

1. Subcortical Unit   The subcortical unit of the brain is functionally specialized for the "maintenance of . . . [an] *optimal level of cortical tone . . .* essential for the organized course of mental activity" (Luria, 1973, p. 45). This functional concept or "activating system" (Brodal, 1981), anatomically is associated with the brain stem reticular formation, which is a group of cells within the regions of the medulla, pons, and midbrain. The morphological features of this functional unit implicate its role in all conscious and autonomic activity (Brodal, 1981; Luria, 1973), and support the notion that it is involved in the activation, inhibition, and regulation of the central nervous system. Its system of projection fibers serves to interconnect this unit of the brain with more rostral (as well as caudal) neuroanatomical structures, again attesting to its mutual interdependence with the cortical units of the brain.

2. Posterior Cortical Unit   The posterior cortical unit of the brain functionally is specialized for the reception, analysis, and storage of information. The neuroanatomical substrate for this functional unit is the cortex (i.e., a six-layered structure that occupies the surfaces of the cerebral hemispheres) of the parietal, occipital, and temporal lobes. Within this functional unit is a hierarchy of zones: primary, secondary, and tertiary.

There is one primary zone within each of the three lobes of this unit, and each primary zone essentially is surrounded by a secondary zone. The major tertiary zone within this unit is demarcated by the area where all three lobes overlap, a region that corresponds to the angular gyrus. The functions of the primary and secondary zones within each lobe are limited to a particular modality. For example, zones within the parietal lobe functionally are specialized for the reception, analysis, and storage of kinesthetic and somat-osensory information; zones within the occipital lobe for visual information; and zones within the temporal lobe for auditory information. In contrast, the tertiary zones function to synthesize this information across modalities.

More specifically, the primary (or projection) zones discriminate among stimuli and influence sensory reception to ensure optimal perception. The secondary (or association, gnostic) zones, in contrast, are relatively more integrative and are adapted to relaying afferent impulses to tertiary zones for further synthesis. Thus, lesions of the primary zones frequently result in specific sensory deficits, whereas lesions of the secondary zones are likely to result in disorganized perceptions within and among complex groups of unimodal stimuli (Luria, 1970). Finally, the tertiary zones functionally are specialized to integrate stimuli across modalities. According to Luria (1973), this integrative function is associated with thinking abstractly and with memorizing information, both of which are cognitive processes that include converting successive stimuli into simultaneously processed groups. Thus, lesions of the tertiary zones of this unit are likely to disrupt simultaneous information processing (Luria, 1980), a holistic kind of problem solving.

3. Anterior Cortical Unit   The anterior cortical unit also comprises three hierarchical zones. In this case, however, it is the tertiary zones that guide the functions of the secondary and primary zones, rather than vice versa as with the posterior cortical unit. The tertiary zones of the anterior cortical unit of the brain functionally are specialized to "play a decisive role in the formation of intentions and programmes, and the regulation and verification of the most complex forms of human behaviour" (Luria, 1973, p. 84). Such behaviors include speech and higher order information processing, which are partially characterized by the successive synthesis of information and maintaining an appropriate set in order to achieve a future goal (Luria, 1973; Shallice, 1982). These behaviors, or executive functions, subserve strategic planning, impulse control, and organized search, as well as flexibility of thought and action (Welsh, Pennington, & Groisser, 1991). Secondary zones functionally are specialized to prepare motor programs and to organize movement, whereas the primary zones functionally are specialized to execute the most basic elements of motoric activity. The neuroanatomical substrate for this functional unit is the cortex of the frontal lobes.

4. Luria-Das Model   The information-processing aspects of Luria's neuropsychological model extensively were elaborated by Das, Kirby, and Jarman (1975, 1979), and soon became known as the "Luria-Das model." The initial research primarily emphasized simultaneous and successive information processing (e.g., Das, 1972, 1973), but later research examined aspects of planning as well (e.g., Ashman & Das, 1980). Thus, this model essentially has targeted the two cortical functional units of Luria's theory for its primary foci and has attempted to operationalize the cognitive processes putatively subserved by those units. The other major functional unit of Luria's theory, that is, the subcortical unit that addresses arousal functions, also has been included in confirmatory factor-analytic investigations (Naglieri, Das, Stevens, & Ledbetter, 1991).

The theoretical basis of the Luria-Das model subsequently was elaborated and significant empirical support now has begun to be established. The elaborated model, called Planning Arousal Simultaneous Successive (PASS) (Naglieri & Das, 1990), is consistent with Luria's conceptualization and assumes that, although each functional unit participates in all cognitive tasks, each functional unit also contributes a specific cognitive function to its associated information processing (Naglieri, Das, & Jarman, 1990). Naglieri and Das (1990) cogently argued that the subcortical functional unit of the brain is a prerequisite for information processing because of its role in maintaining proper levels of arousal (or "cortical tone"). The cortical functional units of the brain, of course, also clearly are seen as central to information processing, given the relationship of simultaneous and successive coding processes to the acquisition, storage, and retrieval of information. Finally, the tertiary (or prefrontal) zone, in particular, of the anterior cortical functional unit is implicated in the application of these coding processes for the efficient planning and verification of behavior. Tasks have been identified to operationalize many of these concepts, and their interactive yet distinct nature has been described (Naglieri & Das, 1990). For example, in a sentence repetition task, the subcortical unit would be involved in attending to the words; the posterior cortical unit would be involved in receiving the words, processing them, and storing them; and the anterior cortical unit would be involved in planning the appropriate response.

Although each functional unit necessarily is involved in all cognitive tasks, different functional units will dominate depending on the nature of those tasks and their information-processing requirements. The subcortical unit, for example, will dominate on tasks that require an individual to be alert to a particular stimulus, such as a discrimination task where one stimulus needs to be chosen from an array. The second functional unit will dominate when the integration of different components of a task is required to solve a problem (Naglieri & Das, 1990). Simultaneous information processing occurs when stimuli are grouped in terms of their relatedness. Here, elements of every stimulus are interrelated to other elements, and the creation of groups from separate elements is achieved using information from any modality across the primary, secondary, and tertiary zones. In contrast to simultaneous information processing, successive information processing occurs when stimuli are arranged in a linear fashion, such as in a digit-span task (Naglieri & Das, 1990). The anterior cortical unit will dominate on tasks requiring decision making about how to solve a problem; executing a problem-solving strategy; accessing attentional, simultaneous and successive processes; monitoring the effectiveness of the strategy; and, if necessary, making appropriate modifications (Naglieri & Das, 1990). Solving a maze problem provides one example because first a course of action must be planned, and, if that action is not successful, then the plan must be modified. A variety of these kinds of executive function tasks now have been described (e.g., Gnys & Willis, 1991).

C. Tridimensional Models

Only recently have tridimensional neuropsychological models of information processing begun to be elaborated. Perhaps this is due, in part, to a preoccupation with the saggital plane that has pervaded much of American neuropsychology. Satz, Strauss, and Whitaker (1990) described a tridimensional model, however, that emphasizes the coronal and horizontal planes as well as the saggital plane of the nervous system. Actually, although much of Luria's work was devoted to understanding cognitive neuropsychology from the perspective of three functional units (addressing primarily horizontal and coronal planes) he, too, addressed all three of Hughlings Jackson's Cartesian coordinates through his law of progressive lateralization of function with its implications for the ontogeny of cerebral hemispheric specialization (Luria, 1973). For example, Luria postulated that functions subserved by the posterior cortical functional unit of the brain progressively transfer with development from primary to secondary to tertiary zones. Through this process, the left cerebral hemisphere typically becomes dominant for speech functions and the right remains nondominant.

Progressive Lateralization and Equipotentiality   A construct related to the law of progressive lateralization is the concept of equipotentiality. The concept of equipotentiality was described by Lashley (1921) to designate the capacities of an intact functional area to carry out the functions that are lost by destruction of the whole, and this concept was used to explain recovery of function from brain lesions. A major premise of both progressive lateralization and equipotentiality, of course, is that cerebral specializations for particular functions may change during the life span. Satz et al. (1990) reexamined the progressive lateralization and equipotentiality hypotheses, both of which have been rejected as nonviable by many neuroscientists for a number of years now. In so doing, they advanced a truly tridimensional neuropsychological model of information processing.

There are abundant anatomical data (Wada, Clarke, & Hamm, 1975; Witelson & Pallie, 1973), physiological data (Davis & Wada, 1977; Gardiner & Walter, 1977; Molfese & Molfese, 1986), and behavioral data (Entus, 1977; Glanville, Best, & Levenson, 1977; Segalowitz & Chapman, 1980) that support the idea that one cerebral hemisphere relative to the other does not become progressively more specialized with development, but instead appears functionally specialized, at least for some processes, at a very young age. This currently well-accepted viewpoint, of course, is incongruent with that expressed by Lenneberg (1967), who considered cerebral dominance as an emergent phenomenon that develops concomitantly with progressive brain differentiation and functional organization. Satz et al. (1990), however, asserted that Lenneberg's hypothesis rests on a misinterpretation of the concept of equipotentiality. Here, equipotentiality is equated with an initial state of bilateral functional symmetry, a concept that is noticeably absent from Lashley's (1921) use of the construct to explain recovery from brain lesions. Instead, Lashley's conception of equipotentiality refers both to the capacity of structures within the damaged hemisphere (i.e., the coronal plane) as well as the opposite, intact hemisphere (i.e., the saggital plane) to subserve a recovered function. As such, this view also clearly acknowledges the coronal plane of the developing brain.

Satz et al. (1990) further emphasized that the concept of equipotentiality, used in the context of the ontogeny of hemispheric specialization, refers to the capacity of intact structures in the hemisphere ipsilateral as well as contralateral to the lesion. Indeed, evidence from studies of aphasia onset following temporal lobectomy and speech arrest following unilateral carotid amytal injections (Penfield & Roberts, 1959; Rasmussen & Milner, 1977; Satz, Strauss, Wada, & Orsini, 1988) show that whether cerebral reorganization following injury to the dominant speech hemisphere is interhemispheric or intrahemispheric may be moderated by age-related (e.g., developmental) factors. Moreover, anatomical studies of postnatal cortex (Campbell & Whitaker, 1986; Whitaker, 1976; Whitaker, Bub, & Leventer, 1981) and clinical studies of childhood aphasia (Brown & Grober, 1983; Brown & Hecaen, 1976) suggest that not all structures within the dominant hemisphere reach functional maturity at the same time; rather, a developmental process of increasing regional specialization may occur.

A major problem, according to Satz et al. (1990), then, is that the concept of initial equipotentiality of the hemispheres, which is assumed in the progressive lateralization hypothesis, inappropriately has been equated with initial functional bisymmetry between the hemispheres. When equipotentiality is considered more appropriately, it becomes clear that the development of hemispheric specialization can be conceptualized as an evolving, dynamic process that includes saggital, horizontal, and coronal planes. In so doing, Satz et al. (1990) challenged the current view of developmental invariance for cerebral hemispheric specialization (e.g., Kinsbourne & Hiscock, 1977) and, perhaps more importantly, elaborated one of the truly tridimensional neuropsychological models of information processing.

II. IMPLICATIONS FOR INDIVIDUAL DIFFERENCES

Two principal implications to these neuropsychological models of information processing address the psychoeducational issues of assessment and intervention. All of the selected models are relevant to both of these issues, but the bidimensional models perhaps more clearly address assessment, whereas the unidimensional and the tridimensional models perhaps more clearly address intervention. Two examples are suggested as illustrations.

A. Assessment

In terms of assessment, the PASS model proffered by Naglieri and Das (1990) provides a clear framework for evaluating cognitive aptitudes from the bidimensional neuropsychological perspective introduced by Luria (1970). Given their relationship to assessment, both of these models explicitly acknowledge individual differences in information processing. In fact, Luria (1980) reported that the component operations that subserve information processing do not mature independently, but instead result from an individual's unique interactions within the environment. Similarly, Das et al. (1979) reported that cultural factors influence coding preferences for cognitive tasks. The PASS model may prove to be especially promising for psychoeducational assessment, given its psychometric properties, its developmental focus, and its comprehensive perspective on information processing. Indeed, some research suggests possible utility for the assessment of clinical groups (e.g., Hurt & Naglieri, 1992; Reardon & Naglieri, 1992).

Despite Luria's (1970, 1973, 1980) major theoretical contribution to the PASS model, however, Luria's own model for assessment differs from the PASS model in an important way. For example, in Luria's model, each case is approached as an individualized experiment, and component operations of cognitive tasks are elucidated through a detailed analysis of the syndrome. This syndrome analysis is used to investigate the neuropsychological organization of information processing for an individual. In contrast, the PASS model was developed through a series of factor analyses that elucidate similar information processing styles among groups. Thus, in the former model, individual differences have been used to identify deficits in information processing, and in the latter model, individual differences have been used to identify reliable forms of information-processing styles (Das & Varnhagen, 1986).

B. Intervention

The unidimensional and tridimensional models provide some direction in terms of intervention. For example, the tridimensional model of Satz et al. (1990) suggests implications for the reorganization of speech function following injury to the left cerebral hemisphere in children. Further, particular psychoeducational treatment suggestions may follow from efficiency versus interhemispheric-transfer explanations for afferent models of cognitive processing. Here, one hypothesis is that for individuals with lateralized brain dysfunctions, remedial approaches might be selected for information-processing tasks that conform to efficiency explanations, whereas circumvention approaches might be preferred for information-processing tasks that conform to interhemispheric-transfer explanations.

As previously noted, efficiency explanations suggest that although the cerebral hemispheres do not execute the component operations for a cognitive task with equivalent efficiency, each hemisphere is capable of executing those operations. Thus, it is reasonable to predict that a unilateral impairment would not preclude the execution of those operations, even when the impaired hemisphere is more efficient than the intact hemisphere for a particular operation. In contrast, interhemispheric-transfer explanations suggest that at least one component operation required by a cognitive task only can be executed by one particular hemisphere. In this case, therefore, a unilateral impairment would preclude the execution of the operation. Thus, an intervention approach that circumvents, rather than remediates, that operation would be required. Current research on the component operations that, for example, subserve reading and research on differential treatments for subtypes of learning disabilities (e.g., Hooper & Willis, 1989; Hynd, 1986), clearly are relevant to these kinds of hypotheses.

III. CONCLUDING REMARKS: SIMILARITIES AND DIFFERENCES AMONG MODELS

In conclusion, there are at least two major differences and two major similarities among the neuropsychological models of information processing selected for illustration. In terms of their differences, first it is clear that the models emphasize different spatial dimensions. For example, the identified unidimensional models, both afferent and efferent, primarily focus on the saggital plane because of their emphasis on the concept of cerebral hemispheric specialization. These models nicely complement the bidimensional models that were introduced because the latter primarily focus on the horizontal plane and on the coronal plane. Indeed, the one tridimensional model that was identified draws heavily from research that was used to support both the unidimensional models (e.g., Kinsbourne & Hiscock, 1983b) as well as the bidimensional models (e.g., Luria, 1973).

Second, although all of the models identified have been supported empirically at each level of analysis, particular models have emphasized particular levels of analysis. For example, the unidimensional models of cerebral hemispheric specialization have been supported at the anatomical and physiological levels, but the majority of their support has been provided at the behavioral level from research on perceptual asymmetries. In contrast, the bidimensional model of Luria (1970, 1973, 1980) primarily relies on research from the anatomical and behavioral levels, whereas the tridimensional model of Satz et al. (1990) requires a synthesis of research from all three levels of analysis. Perhaps it is unsurprising that as models become more complex in terms of spatial dimensions, they rely on more extensive integration of research from anatomical, physiological, and behavioral levels of analysis for their empirical support.

In terms of similarities, first it is clear that a major premise of all the identified models is the assumption that information processing is mediated by networks of component operations, that is, by functional systems. All models explicitly acknowledge that the human central nervous system is highly specialized for function, and that the science of neuropsychology largely is based on the premise that classes of behavior (covert and overt) involve differential neurological substrata. Here, components of neurological systems have been documented for all of the major sensory and motor modalities throughout the life span, arising from prenatal periods of development (Hynd & Willis, 1988). Additionally, these behavioral classes extend beyond sensory and motor functions and include higher order cognitive processes such as attention, information analysis and synthesis, and organization and planning.

Despite the high degree of specialization within the human central nervous system, all of the identified models reject a localizationist perspective on the neuropsychological basis of information processing, which asserts mutual functional exclusivity among circumscribed cerebral regions. Rather, all of these models emphasize the complementary nature of component operations that become integrated to guide a unitary pattern of behavior. Thus, the central nervous system is conceptualized as a differentiated neural network (Kinsbourne, 1982) that provides a biological substrate for information processing. Here, cognitive processing is subserved by systems of interconnected components, that is, by functional systems (Luria, 1980) rather than by circumscribed cerebral regions.

Finally, it is equally clear that all of the identified models are well grounded in empirical support. Because of this, they provide useful frameworks from which to formulate hypotheses and to guide both theoretical and applied research programs. These hypotheses provide a variety of implications for individual differences associated with information processing.

REFERENCES

Ashman, A. F., & Das, J. P. (1980). Relation between planning and simultaneous-successive processing. *Perceptual and Motor Skills, 51,* 371–383.

Brodal, A. (1981). *Neurological anatomy in relation to clinical medicine* (3rd ed.). New York: Oxford University Press.

Brown, J. W., & Grober, E. (1983). Age, sex, and aphasia type. *Journal of Nervous and Mental Disease, 83,* 431–434.

Brown, J. W., & Hecaen, H. (1976). Lateralization and language representation. *Neurology, 26,* 183–189.

Campbell, S., & Whitaker, H. A. (1986). Cortical maturation and developmental neurolinguistics. In J. E. Obrzut & G. W. Hynd (Eds.), *Child neuropsychology: Vol. 1. Theory and research* (pp. 55–72). San Diego: Academic Press.

Das, J. P. (1972). Patterns of cognitive ability in nonretarded and retarded children. *American Journal of Mental Deficiency, 77,* 6–12.

Das, J. P. (1973). Structure of cognitive abilities: Evidence for simultaneous and successive processes. *Journal of Educational Psychology, 65,* 103–108.

Das, J. P., Kirby, J. R., & Jarman, R. F. (1975). Simultaneous and successive synthesis: An alternative model for cognitive abilities. *Psychological Bulletin, 82,* 87–103.

Das, J. P., Kirby, J. R., & Jarman, R. F. (1979). *Simultaneous and successive cognitive processes.* New York: Academic Press.

Das, J. P., & Varnhagen, C. K. (1986). Neuropsychological functioning and cognitive processing. In J. E. Obrzut & G. W. Hynd (Eds.), *Child neuropsychology: Vol. 1. Theory and research* (pp. 117–140). San Diego: Academic Press.

Davidoff, J. B. (1982). Studies with nonverbal stimuli. In J. G. Beaumont (Ed.), *Divided visual field studies of cerebral organization* (pp. 30–55). San Diego: Academic Press.

Davis, A. E., & Wada, J. A. (1977). Hemispheric asymmetry in human infants: Spectral analysis of flash and click evoked potentials. *Brain and Language, 4,* 23–31.

Entus, A. K. (1977). Hemispheric asymmetry in processing of dichotically presented speech and nonspeech stimuli by infants. In S. J. Segalowitz & F. A. Gruber (Eds.), *Language development and neurological theory* (pp. 65–73). San Diego: Academic Press.

Friedman, A., & Polson, M. C. (1981). Hemispheres as independent resource systems: Limited capacity processing and cerebral specialization. *Journal of Experimental Psychology: Human Perception and Performance, 7,* 1031–1058.

Gardiner, M. F., & Walter, D. O. (1977). Evidence of hemispheric specialization from infant EEG. In S. Harnad, R. Doty, L. Goldstein, J. Jayes, & G. Krauthamer (Eds.), *Lateralization in the nervous system* (pp. 481–500). San Diego: Academic Press.

Glanville, B. B., Best, C. T., & Levenson, R. (1977). A cardiac measure of cerebral asymmetries in infant auditory perception. *Developmental Psychology, 13,* 54–59.

Gnys, J. A., & Willis, W. G. (1991). Validation of executive function tasks with young children. *Developmental Neuropsychology, 7,* 487–501.

Hannay, H. J. (Ed.). (1986). *Experimental techniques in human neuropsychology.* New York: Oxford University Press.

Hiscock, M., & Kinsbourne, M. (1987). Specialization of the cerebral hemispheres: Implications for learning. *Journal of Learning Disabilities, 20,* 130–143.

Hooper, S. J., & Willis, W. G. (1989). *Learning disability subtyping: Neuropsychological foundations, conceptual models, and issues in clinical differentiation.* New York: Springer-Verlag.

Hurt, J., & Naglieri, J. A. (1992). Performance of delinquent and nondelinquent males on planning, attention, simultaneous, and successive cognitive processing tasks. *Journal of Clinical Psychology, 48,* 120–127.

Hynd, C. R. (1986). Educational intervention in children with developmental learning disorders. In J. E. Obrzut & G. W. Hynd (Eds.), *Child neuropsychology: Vol. 2. Clinical practice* (pp. 265–297). San Diego: Academic Press.

Hynd, G. W., & Willis, W. G. (1988). *Pediatric neuropsychology.* Orlando, FL: Grune & Stratton.

Jeeves, M., & Baumgartener, G. (Eds.). (1986). Methods in neuropsychology [Special issue]. *Neuropsychologia, 21*(1).

Kandel, E. R. (1985). Cellular mechanisms of learning and the biological basis of individuality. In E. R. Kandel & J. H. Schwartz (Eds.), *Principles of neural science* (2nd ed.) (pp. 816–833). New York: Elsevier.

Kimura, D., & Durnford, M. (1974). Normal studies on the function of the right cerebral hemisphere in vision. In S. J. Diamond & J. G. Beaumond (Eds.), *Hemisphere function in the human brain* (pp. 25–47). New York: Wiley.

Kinsbourne, M. (1970). The cerebral basis of lateral asymmetries in attention. *Acta Psychologica, 33*, 193–201.

Kinsbourne, M. (1973). The control of attention by interaction between the cerebral hemispheres. In S. Kornblum (Ed.), *Attention and performance V* (pp. 239–256). San Diego: Academic Press.

Kinsbourne, M. (1975). The mechanism of hemispheric control of the lateral gradient of attention. In P. M. A. Rabbitt & S. Dornic (Eds.), *Attention and performance V* (pp. 81–97). San Diego: Academic Press.

Kinsbourne, M. (1982). Hemispheric specialization and the growth of human understanding. *American Psychologist, 37*, 411–420.

Kinsbourne, M., & Hicks, R. E. (1978). Functional cerebral space: A model for overflow, transfer and interference effects in human performance: A tutorial review. In J. Requin (Ed.), *Attention and performance VII* (pp. 345–362). Hillsdale, NJ: Erlbaum.

Kinsbourne, M., & Hiscock, M. (1977). Does cerebral dominance develop? In S. J. Segalowitz & F. A. Gruber (Eds.), *Language development and neurological theory* (pp. 171–191). New York: Academic Press.

Kinsbourne, M., & Hiscock, M. (1983a). Asymmetries of dual-task performance. In J. B. Hellige (Ed.), *Cerebral hemisphere asymmetry* (pp. 255–334). New York: Praeger.

Kinsbourne, M., & Hiscock, M. (1983b). The normal and deviant development of functional lateralization of the brain. In P. H. Mussen (Ed.), *Handbook of child psychology: Vol. 2. Infancy and developmental psychobiology* (4th ed.) (pp. 157–280). New York: Wiley.

Lashley, K. S. (1921). Studies of cerebral function in learning. The effects of long continued practice upon cerebral localization. *Journal of Comparative Psychology, 1*, 453–468.

Lenneberg, E. H. (1967). *Biological foundations of language.* New York: Wiley.

Luria, A. R. (1970). Functional organization of the brain. *Scientific American, 222*, 66–78.

Luria, A. R. (1973). *The working brain: An introduction to neuropsychology.* New York: Basic Books.

Luria, A. R. (1980). *Higher cortical functions in man.* New York: Basic Books.

Merola, J. M., & Liederman, J. (1985). Developmental changes in hemispheric independence. *Child Development, 56*, 1184–1194.

Molfese, D. L., & Molfese, V. J. (1986). Psychophysiological indices of early cognitive processes and their relationship to language. In J. E. Obrzut & G. W. Hynd (Eds.), *Child neuropsychology: Vol. 1. Theory and research* (pp. 95–115). San Diego: Academic Press.

Moscovitch, M. (1979). Information processing and the cerebral hemispheres. In M. S. Gazzaniga (Ed.), *Handbook of behavioral neurobiology: Vol. 2. Neuropsychology* (pp. 379–446). New York: Plenum.

Moscovitch, M. (1986). Afferent and efferent models of visual perceptual asymmetries: Theoretical and empirical implications. *Neuropsychologia, 24*, 91–114.

Naglieri, J. A., & Das, J. P. (1990). Planning, attention, simultaneous, and successive (PASS) cognitive processes as a model for intelligence. *Journal of Psychoeducational Assessment, 8*, 303–337.

Naglieri, J. A., Das, J. P., & Jarman, R. F. (1990). Planning, attention, simultaneous, and successive cognitive processes as a model for assessment. *School Psychology Review, 19*, 423–442.

Naglieri, J. A., Das, J. P., Stevens, J. J., & Ledbetter, M. F. (1991). Confirmatory factor analysis of planning, attention, simultaneous, and successive cognitive processing tasks. *Journal of School Psychology, 29*, 1–17.

Penfield, W., & Roberts, L. (1959). *Speech and brain mechanisms.* Princeton, NJ: Princeton University Press.

Rasmussen, T., & Milner, B. (1977). The role of early left-brain injury in determining lateralization of cerebral speech functions. *Annals of the New York Academy of Sciences, 299*, 355–369.

Reardon, S. M., & Naglieri, J. A. (1992). PASS cognitive processing characteristics of normal and ADHD males. *Journal of School Psychology, 30,* 1–13.

Satz, P., Strauss, E., Wada, J., & Orsini, D. L. (1988). Some correlates of intra- and interhemispheric speech organization after left focal brain injury. *Neuropsychologia, 26,* 345–350.

Satz, P., Strauss, E., & Whitaker, H. (1990). The ontogeny of hemispheric specialization: Some old hypotheses revisited. *Brain and Language, 38,* 596–614.

Segalowitz, S. J., & Chapman, J. S. (1980). Cerebral asymmetry for speech in neonates: A behavioral measure. *Brain and Language, 9,* 281–288.

Shallice, T. (1982). Specific impairments of planning. *Royal Society of London, B298,* 199–209.

Wada, J. A., Clarke, R., & Hamm, A. (1975). Cerebral hemispheric asymmetry in humans. *Archives of Neurology, 32,* 239–246.

Welsh, M. C., Pennington, B. F., & Groisser, D. B. (1991). A normative-developmental study of executive functioning: A window on prefrontal function in children. *Developmental Neuropsychology, 7,* 131–149.

Whitaker, H. A. (1976). Neurobiology of language. In E. C. Carterette & M. Friedman (Eds.), *Handbook of perception* (vol. 7) (pp. 429–442). New York: Academic Press.

Whitaker, H. A., Bub, D., & Leventer, S. (1981). Neurolinguistic aspects of language acquisition and bilingualism. In H. Winitz (Ed.), *Native language and foreign language acquisition* (pp. 59–74). New York: New York Academy of Sciences.

Witelson, S. F., & Pallie, W. (1973). Left hemisphere specialization for language in the new-born: Neuroanatomical evidence of asymmetry. *Brain, 96,* 641–646.

6

A Framework for Understanding Individual Differences in Memory
Knowledge–Strategy Interactions

Mitchell Rabinowitz and Daniel W. Kee

Much of the research and theorizing within cognitive science is based on the metaphor that, like a computer, the mind is a general information-processing device. This metaphor implies that both the mind and computer perceive information from an input device and, through a series of processing steps, evaluate and transform the information for understanding and/or output. The adoption of this metaphor has led to the investigation of information-processing systems such as computers.

Two basic premises underlie these types of information-processing systems: (1) the systems operate on physical symbols (information), and (2) these operations are defined as the manipulation of symbols by rules (see Newell, 1979; Newell & Simon, 1972). The theoretical consequence of adopting these premises has been the attempt to specify the "program," the "software," or the rules that people at different levels of skill or ability use to manipulate representations and how rules develop to account for performance.

This orientation has had a significant impact on research aimed at understanding individual differences in memory performance. Much comparative research has indicated that poorer performance is often associated with problems related to the use of strategy processes; i.e., either optimal strategies are not used or there is inefficient use of the strategies. For example, the research contrasting learning-disabled and non-learning-disabled children consistently shows that learning-disabled children are less efficient or exhibit a deficiency in the use of strategic processes (Rabinowitz & Chi, 1987). Although there appears to be a general acceptance of this relation between strategy use and memory performance, the reasons why we observe such differences in the use of strategies are still being debated.

There are two ways in which differences in strategy use can be character- ized. They are (1) deficient and (2) inefficient use of a strategy. Deficient use implies problems concerning strategy knowledge and strategy choice; that is, people may not be aware that they should use a strategy, or an inappropriate strategy is being used. Inefficient use of a strategy implies that a strategy being used is not being used well.

In this chapter we will propose a possible explanation for variations in strategic processing by addressing the relationship between access to existing knowledge and the use of cognitive strategies. Specifically, we will argue that both the decision to use a strategy and the efficiency with which a strategy can be used are based on a complex interaction with the conceptual knowledge to which the strategy is to be applied. Prior to presenting our approach, how- ever, we will first begin by presenting a more elaborate presentation of what strategies are and in that context also present an alternative orientation.

I. STRATEGIES

At a general level, strategies are often described as procedures that are used as an aid in the performance of a given task. Examples of different strategies abound in the psychological literature as do different definitions. An impor- tant characteristic of strategies is that they are goal-oriented processes that are intentionally invoked (Brown, Bransford, Ferrara, & Campione, 1983; Rabinowitz & Chi, 1987).

Rabinowitz and Chi (1987) have made a relevant distinction between *general-context* and *specific-context* strategies. A general-context strategy is defined as one that is exhibited in situations in which a person chooses to use a strategy primarily on the basis of the general constraints of the task. In this situation, the person makes the decision to use the strategy at the start of the task, that is, before the actual materials are viewed. An example of such a strategy, in the case of studying behavior, might be the use of an outlining strategy, such as to glance over the material to get an outline of what might be discussed. The process of attempting to outline is adopted prior to looking at the material. A specific-context strategy is distinguished from a general- context strategy in that the decision to initiate a specific-context strategy is made in response to a specific, rather than the general, situation. This usually entails noticing similarities, differences, or gaps in knowledge. For example, a student might notice while studying, that information from two different paragraphs appears to be similar and as a consequence, engage in additional processing to further explore the similarity.

Given this distinction, how can we then account for individual differences in strategy use? One approach that has developed to address this question can be characterized as the "metacognitive" approach (Brown et al., 1983; Lodico, Ghatala, Levin, Pressley, & Bell, 1983; Pressley, Borkowski, & Schneider, 1987). Underlying this approach is the belief that strategy use involves more than just knowledge about a strategy; it also involves the ability to coordinate and plan the execution of a strategy, and the additional knowledge of when and why a strategy should be used is necessary. Proponents of this view assert that the "good strategy user" is one that is competent in these metacognitive skills. Conversely, poor strategy use is related to poor metacognitive skills (Pressley et al., 1987). Consequently, there has been a sizable amount of research conducted attempting to relate metacognitive skills with strategy use (see Schneider & Pressley, 1989).

In reference to a general context strategy, the decision to use such a strategy is made prior to looking at the specific nature of the materials. An important prerequisite for the use of such a strategy, then, is metacognitive knowledge—knowledge of the general aims of the task and knowledge of which strategies might be applicable. In this context, the specific information presented in the materials, or how that information is related to prior knowledge, cannot influence the decision to use the strategy.

However, the decision to use a specific-context strategy is not made at the start of the task. Subjects use a strategy only in response to certain specific situations, and the ability to notice these specific situations is *based* on prior knowledge. Thus, although the exhibition of general-context strategies depends primarily on metacognitive skills, the exhibition of specific-context strategies depends to a large extent on domain-specific knowledge.

The proposal that strategic processing is dependent on the availability of relevant knowledge is acknowledged, at least to some extent, by practically everyone conducting research pertaining to strategic processing. Unfortunately, knowledge is often only considered in an all-or-none manner with the distinction being between knowledge that is available and information that is unavailable because of lack of knowledge. With such a dichotomy, the role that knowledge can play in accounting for variations in strategic processing becomes minimal and, thus, variations in the observance of strategies are often considered to be independent of the factor of conceptual knowledge. This leads one to inappropriately emphasize the metacognitive component.

However, a distinction needs to be made between the availability and the accessibility of knowledge. The position we will take is that conceptual knowledge can affect strategy use because the application of a given strategy is always interacting with the accessibility of relevant knowledge. The issue of access to semantic knowledge has typically been discussed in reference to the ability to flexibly retrieve information in a variety of contexts (Brown & Campione, 1981). In this chapter, we refer to access to knowledge in a more limited sense, that is, as the speed with which semantic knowledge can be accessed in a clearly defined context. All information is not equally accessible.

II. REPRESENTING KNOWLEDGE

This variation in accessibility to knowledge is implied by representational models of conceptual knowledge. Conceptual knowledge is often described in terms of a specific representational system—a spreading-activation memory system (Anderson & Bower, 1973; Anderson & Pirolli, 1984; Collins & Quillian, 1969; Norman & Rumelhart, 1975). Some common assumptions associated with such a representational system are that concepts are represented as nodes that are interconnected by associative links that can vary in strength. When an item is encountered, the corresponding concept in memory is activated. Activation then spreads from that concept across associative links to related concepts. The amount of activation that spreads depends on the strength of the associative links between concepts, with the stronger association leading to stronger activation.

This spread of activation within the knowledge base is thought to occur automatically and not be under the conscious control of the learner. Thus, whereas strategic behavior might be characterized as goal-oriented, the automatic spread of activation cannot. Whether activation will spread from concept to concept in memory is determined by the level of initial activation and the existence of associative links, and not necessarily the intentions or metacognitive skills of the person. Any given concept can initially become activated through a variety of goal-oriented (strategic) or non-goal-oriented (spread of activation) processes. Once the concept becomes active, however, it will spread activation to its neighbors regardless of the goal-oriented behavior going on. Thus, the spread of activation affects the activation levels of concepts in a manner that is independent of any specific goal-oriented or strategic behavior.

The work on spreading activation models was developed as part of the metaphor "the mind is like a computer." As stated earlier, one of the implications of the procedural perspective is that the rules need to operate on physical symbols and these symbols have to be stored in some memory location. Retrieval involves going to where it was stored and finding it. This orientation led to the investigation of the content and structure of declarative knowledge.

Early models (e.g., Collins & Loftus, 1975) only specified excitatory links interconnecting nodes. Excitatory links increase the level of activation of a node's neighbors. Thus, when a node becomes active, it activates related nodes, some of which might be relevant to the task at hand, many of which are not. In such a system, there was no way for a spreading-activation system to make "intelligent" decisions regarding knowledge activation. Some other process (rules, procedures) is needed to accomplish the intelligent aspects of cognition. For this reason, the role that spreading-activation systems play in accounting for performance has been thought to be minimal.

However, recently, conceptions of network models have changed due to the adoption of the metaphor "the mind is like the brain" (e.g., connectionist models, Feldman & Ballard, 1982; Rumelhart & McClelland, 1986). In these models, the manner in which neurons process information is seen to be a good model for understanding spreading activation. The neuron processes information by the transmission of neurotransmitters from the axon terminal of one neuron, across the synapse, to the dendrite of another neuron. These neurotransmitters act to change the polarization, or electrical potential, of the receptor neuron. Depending on the neurotransmitter, the electrical potential of the receptor neuron can either increase or decrease. Thus, the neurotransmitter can either be excitatory or inhibitory. (See Crick & Asanuma, 1986, for an in depth discussion of the relation between neuronal architecture and connectionist models.)

The important point for our purposes here is that the adoption of the metaphor "the mind is like a brain" led investigators to investigate network systems that entail inhibitory, as well as excitatory links. The inclusion of inhibitory links changes the nature of spreading-activation systems such that they become *constraint satisfaction* models. Rather than allowing all the neighbors to become activated, inhibitory links provide a mechanism whereby context constrains activation. A number of researchers have posited that rules or procedures can be implicitly represented within the associative links (Rabinowitz, Lesgold, & Berardi, 1988; Rumelhart & McClelland, 1986). This permits spreading-activation systems to make intelligent choices regarding knowledge activation and argues that such a system can play a much more important role in determining performance characteristics.

III. KNOWLEDGE AND STRATEGY INTERACTIONS

In the preceding section we suggested that the relative access to knowledge can have a significant effect on strategy use. In this section, we will present experimental evidence to support this position. This section will illustrate how access to knowledge affects strategy efficacy, and an account for this effect will be offered based on the notion of mental effort. For this purpose, the use of two different memory strategies, categorization and elaboration, will be illustrated and discussed.

A. Categorization

One task that has been used to study individual differences in memory is the task where people are asked to memorize a list of categorizable words. With this task, a good strategy to use is to attempt to group the items by categories during encoding and to use that organizational scheme as an aid during recall. Previous research has shown that, in most situations, people who categorize, recall more items than people who do not (Schneider & Pressley, 1989; but see Rabinowitz, 1991 for a situation where this does not hold.)

The categorization strategy, however, is never used in isolation; it must be applied to specific category exemplars, and exemplars of a given category vary in terms of the speed with which they can be accessed in the context of a category. Some exemplars are more typical of a category than others (for animals, *cat* and *dog* vs. *mule* and *sheep*), and this variation in typicality seems to affect the accessibility of the information. Rosch (1973) found that adults are faster at verifying that a highly typical exemplar is a member of a category than they are at verifying a less typical exemplar. Furthermore, Mervis, Catlin, and Rosch (1976) indicated that adults tend to list highly typical exemplars more often than less typical exemplars when asked to generate exemplars from given superordinate categories. Thus, the categorical information associated with highly typical exemplars seems to be more accessible than is that associated with less typical members.

One hypothesis that was suggested in the earlier section of this chapter was that variation in accessibility to knowledge will affect the efficiency with which knowledge can be used and as a result the effectiveness of the strategy. Rabinowitz (1984) provided support for this hypothesis. In that study, second and fifth graders were asked to memorize two categorizable lists of words. One list consisted of items that were highly typical of their respective categories (e.g., dog for animals); the other consisted of items that were less typical (e.g., goat). In addition, subjects were given one of three types of memory instructions: standard free recall, repetition, or categorization instructions. In the categorization condition, subjects were informed of the categorical nature of the list and were instructed to try to group the items by categories. As expected, performance was better for subjects in the categorization condition than for subjects in the repetition and standard free-recall conditions. However, the interesting finding is that the subjects in the categorization condition benefited more with the highly typical list than with the less typical list. The effectiveness of the strategy depended on the typicality of the exemplars to their related superordinates. A number of other studies have also shown that recall and clustering are greater when college students are asked to study a list consisting of highly typical exemplars than when asked to study lists consisting of less typical exemplars (Bjorklund & Buchanan, 1989; Greenberg & Bjorklund, 1981; Rabinowitz, 1988, 1991; Rabinowitz, Freeman, & Cohen, 1992).

One consequence of this result is that these variations in typicality might also affect subjects' perception of the categorization strategy. Subjects engaging in an initial experience of applying the strategy to a list of highly typical exemplars should be able to recall more items and, as a consequence, be more likely to perceive the strategy as effective and easy to use than would those who applied the strategy to a list of less typical items. Using subjects' self-report data, Rabinowitz (1988) showed that subjects rated that the categorization strategy was more difficult to use, required more effort, and was less useful with a list of low-typicality items than with a list of high-typicality items. (See also Bjorklund & Harnishfeger, 1987; and Bjorklund, Coyle, & Gaultney, 1992, for a related discussion on effort and categorization).

B. Elaboration

A second task that has been used to investigate differences in strategy use is the paired-associate task. With this task, people are presented with pairs of items—often unrelated concrete nouns—on a study trial. They are told to memorize the pair so that on a cued-recall test they can recall one member of the pair when presented with its mate. Research has demonstrated that a good strategy to use with this task is an elaboration strategy and that individual and developmental differences in performance have been shown to be related to the use of this strategy (Pressley, 1982). Use of the elaboration strategy requires the subject to create a semantic event that includes the to-be-learned items in a direct interaction (see Rohwer, 1973). For example, an effective elaboration for the pair arrow—glasses is the *arrow* smashes the *glasses.*

Rohwer, Rabinowitz, and Dronkers (1982) proposed that some types of learners (e.g., younger as opposed to older) are less likely to have fast access to *events* in their knowledge base that include or overlap the to-be-learned items. Thus, for these learners elaboration of unrelated nouns is most likely a multistep procedure consisting of (1) location of possible events for each member of the pair; (2) selection of events that could include *both* members; and (3) a final transformation of the events so that both pair members can be included. In contrast, elaboration by more mature learners is less complicated because they need only locate events in their repertoires that overlap the to-be-learned items.

Accessibility to relevant information for elaboration has been experimentally manipulated using materials consisting of "accessible" and "inaccessible" word pairs (Rohwer et al., 1982). Accessible (e.g., fish—seaweed) involve items that occur naturally and share natural events. Inaccessible pairs (e.g., fish—napkin) also occur naturally, but do not share natural events. Studies using these materials have found that age differences in performance are smaller when subjects are instructed to elaborate and are presented accessible compared to inaccessible pairs (Kee & Davies, 1991; Rohwer et al., 1982).

Recent research concerning the use of elaboration has focused on the impact of variations in speed and efficiency of access to strategy-relevant event information in the knowledge base. For example, studies by Kee and Davies (1988, 1991) have revealed that the developmental increase in elaborative efficacy is associated with a corresponding decrease in the mental effort required to elaborate pairs. In these studies, dual-task methods were used to estimate mental-effort expenditure. Subjects completed a series of 10 s finger tapping trials. On some of these trials subjects were also asked to use elaboration in the learning of unrelated nouns (dual task). Finger tapping interference (percentage of baseline change in finger tapping performance) provides the index of mental-effort expenditure (see Guttentag, 1989b).

The character of the noun-pair lists to be learned have been manipulated in these studies to determine if the developmental differences in mental effort could be attributed to knowledge-base access differences. For example, Kee and Davies (1990, 1991) required subjects to elaborate accessible and inaccessible pairs under dual-task conditions. Kee and Davies (1990) reported that the mental effort expended during the creation of sentence elaborations by fifth-grade children was greater for inaccessible pairs than accessible pairs. Furthermore, this accessibility difference was not observed when subjects were provided with the pair members in a sentence elaboration. A developmental study by Kee and Davies (1991) compared the performance of third- vs. seventh-grade children. A List by Grade interaction in finger tapping interference was observed indicating accessibility list differences for the third-grade children, but not the seventh-grade children. The absence of an accessibility difference at the seventh grade was attributed to attenuation of the strength of the list manipulation (accessible vs. inaccessible) due to age-related increases in the availability of relevant semantic events for the elaboration of the nouns.

The preceding examples focused on developmental differences. In contrast, a study by Kee, Yokoi, and Cafaro (1991) reveals that gender differences in knowledge-base accessibility can also affect elaborative strategy use with second-grade children. Using a dual-task procedure, performance for three different types of noun-pair lists was examined: *masculine* pairs, *feminine* pairs, and *mixed* pairs. Masculine pairs consisted of nouns that shared a masculine connotation (e.g., airplane—hammer), while feminine pairs consisted of nouns that shared a feminine connotation (e.g., actress—flower). Mixed pairs consisted of one noun from each category. The 18 nouns used were drawn from previous studies and the experimental lists used were constructed so that nouns appeared in both the mixed and masculine or feminine lists. Counterbalancing across subjects assured that each noun appeared only once in each subject's mixed-list of pairs. Children were tested using dual-task procedures: finger tapping and elaboration. Subjects tapped microkeys as quickly as possible with the index fingers of both hands. On some trials they only tapped, while on others they also constructed a sentence elaboration for the pair.

The results from this study indicated that overall greater mental effort was required for children to generate elaborations for mixed pairs than for masculine and feminine pairs. For boys, sentence elaborations were less effortful for masculine pairs in contrast to feminine pairs and mixed pairs, respectively. In contrast, for girls sentence elaborations were less effortful for feminine pairs in contrast to mixed pairs. No difference in finger tapping interference was observed between masculine vs. feminine pairs.

Recall that elaboration of nouns can be viewed as a multistep procedure that includes location of possible events for each noun; selection of events that can include both nouns; and a final transformation of events so that both nouns can be included. Fewer steps are involved if semantic events can be activated that already include the nouns. Based on this framework, the findings from Kee et al. indicate increased knowledge-base access for elaboration of nouns that share a consistent gender connotation. Furthermore, differential access to relevant semantic events was suggested for the elaboration of feminine pairs by boys and girls.

C. Spontaneous Strategy Use by Knowledge-Base Interaction

Learners' successful use of elaboration requires that they have direct access to *events* in their knowledge base that include or overlap the to-be-learned items. In the absence of direct access, the use of elaboration is *less* resource efficient because of the additional steps that are required to create a direct interaction for the pair members. An early study by Beuhring and Kee (1987) showed that when subjects are not instructed to generate elaborations, they often produce other forms of relational associations during the associative learning of noun-pairs. Two major categories identified in their children's speak-aloud protocols were Elaboration and Other Associative Strategies. Elaboration was characterized by a *direct* interaction between pair members (e.g., the *fish* swam in the *seaweed*), while Other Associative Strategies consisted of descriptions that offered associations other than a direct interaction. Examples include relational associations based on a common attribute, a negative relationship, or a transformation of a noun to another part of speech (for a complete description refer to Beuhring & Kee, 1987).

In order to more fully understand the relationship between knowledge-base access and *spontaneous* strategy use, Kee and Guttentag (1991) asked fourth/fifth-and seventh/eighth-grade children to learn *accessible* and *inaccessible* pairs under instructions to generate sentences aloud that would help them remember the items. (Note, subjects were *not* told to construct elaborations for the pairs as in previously discussed studies). The time required by subjects to generate sentences for each pair was recorded (pair offset to sentence description onset) and these sentences were subsequently classified into the categories developed by Beuhring and Kee (1987). The response time and recall analyses conducted by Kee and Guttentag indicated a trade-off between the efficacy of knowledge-base access (indexed by speed) and recall for pairs encoded by Other Associative Strategies, but not Elaboration. That is, for Other Associative Strategies, the additional time required to generate a sentence for inaccessible pairs (in contrast to accessible pairs) was associated with a recall penalty. In contrast, Elaboration—by definition a strategy that reflects more direct knowledge-base access—was not associated with such a trade-off. Again, this finding accords well with the notion that faster processing reflects more resource-efficient processing, resulting in greater availability of mental resources for other task-relevant processing (Bjorklund, Muir-Broaddus, & Schneider, 1990; Rabinowitz & Chi, 1987).

D. Knowledge, Strategies, and Cerebral Hemisphere Specialization

The previous section provided illustrations of how access to relevant knowledge can affect strategy efficacy. The concepts of mental resources and effort were introduced to account for the observed relationships. Current research suggests that independent pools of mental resources are associated with the left and right cerebral hemispheres (see Friedman, Polson, Dafoe, & Gaskill, 1982). Furthermore, Hellige (1993) has suggested that we consider the hemispheres as "two general information processing subsystems with different biases and abilities." Dual-task procedures—similar to the finger tapping task used in the mental effort studies of knowledge access—can be used to examine the relative involvement of the left vs. right cerebral hemispheres (see Hellige & Kee, 1990; Kinsbourne & Hiscock, 1983). That is, observed asymmetric finger tapping interference produced by concurrent cognitive activities can implicate hemispheric differences because (1) finger tapping of each hand is controlled (programmed) primarily by the contralateral hemisphere and (2) two tasks should interfere more with each other if they compete for mental resources from within the same hemisphere as opposed to two different hemispheres. For example, studies show that more right-hand than left-hand finger tapping interference is observed when subjects are asked to perform verbal concurrent activities such as rhyme recitation, solving anagrams (e.g., Kee, Bathurst & Hellige, 1984) and holding words in short-term memory (e.g., Friedman, Polson, & Dafoe, 1988). This pattern of asymmetric finger tapping interference indicates more left than right hemisphere involvement in the processing of these verbal-language activities.

Some time ago, Paivio and le Linde (1982) hypothesized that differences in the efficacy of verbal vs. imaginal strategies may be correlated with the specializations of the cerebral hemispheres. A dual-task study by Lempert (1989) supports their notion, indicating different patterns of asymmetric interference for visual imagery in contrast to verbal rehearsal of text. Overall, however, little is known about hemispheric involvement for memory strategies and knowledge access. It would not be surprising, however, if asymmetric cerebral hemisphere involvement was uncovered, because research clearly indicates that the cerebral hemispheres are associated with different processing strategies (see Hellige, in press-a). In this regard, the dual-task finger tapping procedure provides an excellent means to explore whether (1) different memory strategies demand more left vs. right hemisphere resources and (2) the extent to which asymmetric involvement reflects an interaction with knowledge access.

IV. A RETURN TO METAPHORS

We began this chapter with a discussion of the metaphor "the mind is like a computer." We argued that the metaphor was important because it provided a framework in which to conceptualize the issue of individual differences in memory. Adoption of this framework has led researchers to emphasize the role that strategic processing plays in accounting for memory differences and, we feel, underemphasizing the role that knowledge might play.

There are two primary reasons why this occurred. First, inherent within the computer metaphor, is the distinction made by Newell (1979) between structure and process. Knowledge structures (the physical symbols) are seen as static, permanent, and objectlike, whereas process is seen as dynamic, transient, and transformationlike. The process–structure distinction defined the relative roles of process and structure, with structure essentially limited to and enabling role. Procedures require certain symbols to be available before the procedures can be applied. Second, while spreading-activation models were actually developed within the context of the computer metaphor, the early models (e.g., Collins & Loftus, 1975) only specified excitatory links interconnecting nodes. Excitatory links increase the level of activation of a node's neighbors. Thus, when a node becomes active, it activates related nodes, some of which might be relevant to the task at hand, many of which are not. There was no way for a spreading activation system to make "intelligent" choices regarding knowledge activation. Thus, some other process (strategy) was needed to accomplish the intelligent aspects of cognition.

The recent emphasis on the role that knowledge might play in accounting for individual differences in memory developed concomitantly with developments in the characterization of network models and the adoption of the consideration of a different metaphor—"the mind is like a brain." This second metaphor is based on conceptions of how neurons process information. The recent work on connectionist models (Feldman & Ballard, 1982; Rumelhart & McClelland, 1986) has suggested that the information-processing mechanisms of neurons might provide a better model for understanding human cognition.

These connectionist models developed from spreading-activation models but they posit the use of inhibitory as well as excitatory links. Inhibitory links serve to decrease the level of activation of a node's neighbors. The inclusion of inhibitory links enables spreading-activation models to become "constraint satisfaction" models. Rather than allowing all the neighbors of a node to become activated, inhibitory links provide a mechanism whereby the context constraints activation. This permits spreading-activation systems to make "intelligent" choices regarding knowledge activation.

The upshot of all this is that research from both the psychological and neuropsychological perspectives needs to explore the interactive relation between knowledge and strategy use. Strategy use clearly is a factor underlying individual differences in memory performance, but the knowledge factor might offer a causal explanation of why there are differences in strategy use. Knowledge should not be conceptualized in an all-or-none fashion but rather activation of knowledge varies on a continuum of accessibility. Research investigating issues of connectivity and accessibility will be a fruitful way to proceed in investigating individual differences in memory.

REFERENCES

Anderson, J. R., & Bower, G. H. (1973). *Human associative memory.* Washington, DC: V. H. Winston.

Anderson, J. R., & Pirolli, P. L. (1984). Spread of activation. *Journal of Experimental Psychology: Learning, Memory, & Cognition, 10,* 791–798.

Beuhring, T., & Kee, D. W. (1987). The relationships between memory knowledge, elaborative strategy use and associative memory performance. *Journal of Experimental Child Psychology, 44,* 377–400.

Bjorklund, D. F., & Buchanan, J. J. (1989). Developmental and knowledge base differences in the acquisition and extension of a memory strategy. *Journal of Experimental Child Psychology, 48,* 451–471.

Bjorklund, D. F., Coyle, T. R., & Gaultney, J. F. (1992). Developmental differences in the acquisition and maintenance of an organizational strategy: Evidence for a utilization deficiency hypothesis. *Journal of Experimental Child Psychology, 54,* 434–448.

Bjorklund, D. F., & Harnishfeger, K. K. (1987). Developmental differences in the mental effort requirements for the use of an organizational strategy in free recall. *Journal of Experimental Child Psychology, 44,* 109–125.

Bjorklund, D. F., Muir-Broaddus, J. E., & Schneider, W. (1990). The role of knowledge in the development of strategies. In D. F. Bjorklund (Ed.), *Children's strategies: Contemporary views of cognitive development* (pp. 93–128). Hillsdale, NJ: Erlbaum.

Brown, A. L., Bransford, J. D., Ferrara, R. A., & Campione, J. C. (1983). Learning, remembering, and understanding. In J. H. Flavell & E. Markman (Eds.), *Handbook of child psychology: Cognitive development, Vol. 3* (pp. 77–166). New York: Wiley.

Brown, A. L., & Campione, J. C. (1981). Inducing flexible thinking: A problem of access. In M. Friedman, J. P. Das, & N. O'Connor (Eds.), *Intelligence and learning* (pp. 515–529). New York: Plenum Press.

Collins, A. M., & Loftus, E. F. (1975). A spreading activation theory of semantic processing. *Psychological Review, 82,* 407–428.

Collins, A. M., & Quillian, M. R. (1969). Retrieval times from semantic memory. *Journal of Verbal Learning and Verbal Behavior, 8,* 240–247.

Crick, F. H. C., & Asanuma, C. (1986). Certain aspects of the anatomy of the cerebral cortex. In J. L. McClelland and D. E. Rumelhart (Eds.), *Parallel distributed processing: Explorations in the microstructure of cognition. Vol. 2. Psychological and biological models* (pp. 333–371). Cambridge, MA: MIT Press.

Feldman, J. A., & Ballard, D. H. (1982). Connectionist models and their properties. *Cognitive Science, 2,* 205–254.

Friedman, A., Polson, M. C., & Dafoe, C. G. (1988). Dividing attention between the hand and the head: performance trade-offs between rapid finger tapping and verbal memory. *Journal of Experimental Psychology: Human Perception and Performance, 14,* 60–68.

Friedman, A., Polson, M. C., Dafoe, C. G., & Gaskill, S. (1982). Dividing attention within and between hemispheres: testing a multiple resources approach to limited-capacity information processing. *Journal of Experimental Psychology: Human Perception and Performance, 8,* 625–650.

Greenberg, M. S., & Bjorklund, D. F. (1981). Category typicality in free recall: Effects of feature overlap or differential category encoding. *Journal of Experimental Psychology: Human Learning and Memory, 7,* 145–147.

Guttentag, R. E. (1989). Age differences in dual-task performance: Procedures, assumptions, and results. *Developmental Review, 9,* 146–170.

Hellige, J. B. (1993). Unity of thought and action: Varieties of interaction between the left and right cerebral hemispheres. *Current Directions in Psychological Science, 2,* 21–23.

Hellige, J. B. (in press-a). Coordinating the different processing biases of the left and right cerebral hemispheres. In F. L. Kitterle (Ed.), *Hemispheric interaction.* Hillsdale, NJ: Erlbaum.

Hellige, J. B., & Kee, D. W. (1990). Concurrent task performance as an indicator of lateralized brain function. In G. R. Hammon (Ed.), *Advances in Psychology* (pp. 635–660). Netherlands: North-Holland.

Kee, D. W., & Davies, L. (1990). Mental effort and elaboration: Effects of accessibility and instruction. *Journal of Experimental Child Psychology, 49,* 264–274.

Kee, D. W., & Davies, L. (1991). Mental effort and elaboration: A developmental analysis of accessibility effects. *Journal of Experimental Child Psychology, 52,* 1–10.

Kee, D. W., & Guttentag, R. E. (1991, April). *Knowledge-base access effects in associative strategy use.* A poster presentation at the biennial meeting of the Society for Research in Child Development, Seattle, WA.

Kee, D. W., Yokoi, L., & Cafaro, T. (1991, April). *Knowledge-base accessibility effects: Boys and girls show different patterns of elaboration effort.* A poster presentation at the biennial meeting of the Society for Research in Child Development, Seattle, WA.

Kee, D. W., Bathurst, K., & Hellige, J. B. (1984). Lateralized interference in finger tapping: Assessment of block design activities. *Neuropsychologia, 22,* 197–203.

Kinsbourne, M., & Hiscock, M. (1983). Asymmetries of dual task performance. In J. B. Hellige (Ed.), *Cerebral hemisphere asymmetry: Method, theory, and application* (pp. 255–334). New York: Praeger.

Lempert, H. (1989). Effect of imaging vs. silently rehearsing sentences on concurrent unimanual tapping: a follow-up. *Neuropsychologia, 27,* 575–580.

Lodico, M. G., Ghatala, E. S., Levin, J. R., Pressley, M., & Bell, J. A. (1983). The effects of strategy-monitoring training on children's selection of effective memory strategies. *Journal of Experimental Child Psychology, 35,* 263–277.

Mervis, C. B., Catlin, J., & Rosch, E. (1976). Relationships among goodness-of-example, category norms, and word frequency. *Bulletin of the Psychonomic Society, 7,* 283–284.

Newell, A. (1979). One last word. In D. T. Tumas & F. Reif (Eds.), *Problem solving and education: Issues in teaching and research* (pp. 175–189). Hillsdale, NJ: Erlbaum.

Newell, A., & Simon, H. A. (1972). *Human problem solving.* Englewood Cliffs, NJ: Prentice Hall.

Norman, D. A., & Rumelhart, D. E. (1975). *Explorations in cognition.* San Francisco: Freeman.

Paivio, A., & le Linde, J. (1982). Imagery, memory, and the brain. *Canadian Journal of Psychology, 36,* 243–272.

Pressley, M. (1982). Elaboration and memory development. *Child Development, 53,* 296–309.

Pressley, M., Borkowski, J. G., & Schneider, W. (1987). Cognitive strategies: Good strategy users coordinate metacognition and knowledge. In R. Vasta & G. Whitehurst (Eds.), *Annals of child development, Vol. 5* (pp. 89–129). Greenwich, CT: JAI Press.

Rabinowitz, M. (1984). The use of categorical organization. Not an all-or-none situation. *Journal of Experimental Child Psychology, 38*, 338–351.

Rabinowitz, M. (1988). On teaching cognitive strategies: The influence of accessibility of conceptual knowledge. *Contemporary Journal of Educational Psychology, 13*, 229–235.

Rabinowitz, M. (1991). Semantic and strategic processing: Independent roles in determining memory performance. *American Journal of Psychology, 104*, 427–437.

Rabinowitz, M., & Chi, M. T. H. (1987). An interactive model of strategic processing. In S. J. Ceci (Ed.), *Handbook of the cognitive, social, and physiological characteristics of learning disabilities, Vol. 2* (pp. 83–102). Hillsdale, NJ: Erlbaum.

Rabinowitz, M., Freeman, K., & Cohen, S. (1992). Use and maintenance of strategies: The influence of accessibility to knowledge. *Journal of Educational Psychology, 84*, 211–218.

Rabinowitz, M., Lesgold, A. M., & Berardi, B. (1988). Modeling task performance: Rule-based and connectionist alternatives. *International Journal of Educational Research, 12*, 35–48.

Rohwer, W. D., Jr. (1973). Elaboration and learning in childhood and adolescence. In H. W. Reese (Ed.), *Advances in child development and behavior, Vol. 8* (pp. 2–59). New York: Academic Press.

Rohwer, W. D., Jr., Rabinowitz, M., & Dronkers, N. F. (1982). Event knowledge, elaborative propensity, and the development of learning proficiency. *Journal of Experimental Child Psychology, 33*, 492–503.

Rosch, E. (1973). On the internal structure of perceptual and semantic categories. In T. E. Moore (Ed.), *Cognitive development and the acquisition of language.* New York: Academic Press.

Rumelhart, D. E., & McClelland, J. L. (Eds.) (1986). *Parallel distributed processing, Vol. 1.* Cambridge, MA: MIT Press.

Schneider, W., & Pressley, M. (1989). *Memory development between 2 and 20.* New York: Springer-Verlag.

Tulving, E., & Donaldson, W. (Eds.) (1972). *Organization of memory.* New York: Academic Press.

Section III
Personality

7

Personality
Biological Foundations

H. J. Eysenck

I. INTRODUCTION

The term personality is used in many different ways, and it would be inappropriate here to discuss in detail these various definitions, and argue about the relative merits of idiographic and nomothetic approaches. Scientific research has largely been based on a conception of personality as being defined and measured in terms of a hierarchical arrangement of traits, intercorrelations between which produce higher order factors, whereas the traits themselves are based on intercorrelations of items, (i.e., specific action tendencies of one kind or another). Traits can be measured by means of self-ratings (questionnaires), ratings by friends or spouses, behavior in miniature situations, reactions in experimental settings, projective tests, and physiological reactions. Some authors have concentrated more on traits than on higher order concepts (e.g., Cattell), while others have emphasized higher order type-concepts (e.g., Eysenck). A detailed account of the arguments, facts, and theories involved has been given elsewhere (H. J. Eysenck and M. W. Eysenck, 1985), and will not be repeated here.

How many superfactors (type concepts) are there, and is there agreement on this point? Eysenck (1991b) has argued that there are three major dimensions of personality, looking at the evidence available at present, and not ruling out the possibility that further dimensions may be added later. These three dimensions are extraversion (E), neuroticism (N), and psychoticism (P); the nature of these type-concepts can be clarified by looking at the traits which, by their intercorrelations, define the types. Figure 1 shows the traits defining E, with introversion of course showing the inverse pattern. The distribution on extraversion–introversion is normal, (i.e., it follows a Gaus-

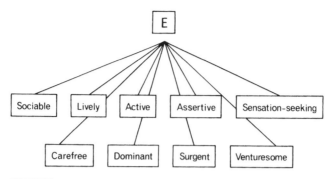

FIGURE 1

Primary traits defining extraversion.

sian curve, with the majority intermediate, and only a few people at either extreme).

Figure 2 does the same service for N. It should be made clear that "N" is not to be identified with "neurosis." N is a *dispositional* variable predisposing a person to develop neurotic symptoms under stress, but a high-N scorer is not *eo ipso* neurotic. Figure 3 shows the scores on an N questionnaire filled in by 1000 neurotics and 1000 normals. It will be seen that 10.6% of normals receive scores above the (arbitrary) dividing line, while 71.4% of neurotics do so. Low scores are obtained by 89.4% of normals, but only by 28.6% of neurotics. Thus a high score gives a 9:1 probability of being neurotic (i.e., under treatment), but 1 in 10 has a high score without being psychiatrically ill (H. J. Eysenck, 1952a). Of course psychiatric diagnosis is far from perfect, but these data do suggest that neuroticism is a *graded,* or *continuous* variable, whereas neurosis may add some more qualitative ingredient, such as a specific breakdown.

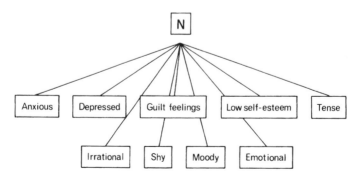

FIGURE 2

Primary traits defining neuroticism.

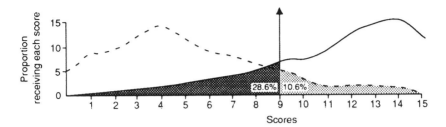

FIGURE 3

Neuroticism scores of diagnosed neurotics (solid line) and normals (evoked line). (From H. J. Eysenck, 1952a.)

The same argument applies to P, shown in Figure 4. Here again the concept is one of a continuous dispositional variable underlying possible psychotic breakdown (Eysenck & Eysenck, 1976). Figure 5 illustrates the concept. The abscissa shows the graded nature of P from psychotic illness on the right to altruistic, socialized, conventional behavior on the left. The curve P_A suggests the probability of people at various points on this scale having a psychotic breakdown; going from left to right this probability increases gradually. This is the so-called diathesis–stress concept; P contributes the diathesis, which, under stress, issues in actual psychotic disorder.

Concepts like P are of course based on a large body of evidence (H. J. Eysenck, 1992). Thus it has been shown that different functional psychoses (e.g., schizophrenia, manic–depressive disorder) are not in fact categorically different but often merge into each other, giving rise to the large intermediate group of schizoaffective disorders (Crow, 1986). Psychotic states are closely related to schizoid behavior, psychopathy, and other traits, and gradually blend into normality. A variety of behavioral, physiological, hormonal, experimental, and other variables, clearly distinguishing between schizophrenics

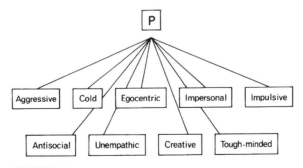

FIGURE 4

Primary traits defining psychoticism.

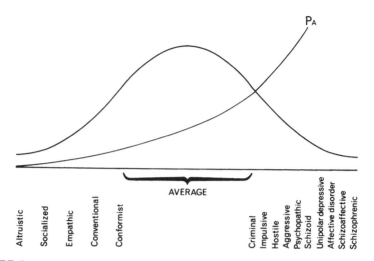

FIGURE 5

Distribution of psychoticism scores (abscissa) and probability of developing psychosis (P_A). (From H. J. Eysenck, 1992.)

and normals, correlate with measures of P both within a normal and also within a psychotic group of subjects (H. J. Eysenck, 1950, 1952b, 1992). It is along these lines that we can support the theoretical concept of P.

The postulation of these three fundamental dimensions finds strong support in a metanalysis by Royce and Powell (1983) of published correlational and factor-analytic studies in this field; they arrive at three superfactors that have a strong similarity to P, E, and N. A further reason for concentrating on these three dimensions is simply that most of the published studies on the biological foundations of personality (physiology, hormones, neurology, enzymes, drug effects, etc.) have been related to these dimensions (Zuckerman, 1991). Furthermore, studies of the biological correlates (causes?) of differential behavior should be based on well-grounded theories, and such theories hardly exist outside the field suggested. This is not to suggest that alternative solutions to the question of a paradigm of personality taxonomy have not been suggested; perhaps the leading contender at the moment is the "Big Five" concept (John, 1990), which adds openness (O), agreeableness (A), and conscientiousness (C) to E and N. I have argued that these three suggested superfactors are in fact primaries (traits), and that, for example A and C are part of the set of traits that define (negatively) the P superfactor (Eysenck, 1991b). Both A and C correlate quite highly (negatively) with P (the multiple R, disattenuated, is −.85!), and such a high correlation would seem to support this view. But however this may be, there are hardly any theories or studies linking O, A, or C with any biological data, and hence they would not find a place in our account anyway.

There are of course problems in the simple hierarchical scheme. Some concepts (e.g., impulsiveness, sensation seeking) can be shown to break down into several subfactors correlating only about .3 with each other yet these are not superfactors but make up the list of primaries on which the superfactors are built. Should we admit one or more intermediate levels between primaries and "type" or superfactors? The point has been discussed by H. J. Eysenck and M. W. Eysenck (1985). Another problem is that primaries do not always define just one superfactor; they may correlate with two, or even all three (e.g., impulsiveness and sensation seeking share variance with P, E, and N!). Simple schematic illustrations like Figures 1, 2, and 4 disregard such complexities, but for the theoretician they must always be borne in mind.

It would be a great mistake to consider work on the biological basis of personality as a mere extension of a taxonomic framework. Cronbach (1957) has drawn attention to the sad division existing between the two disciplines of scientific psychology, the experimental and the correlational, arguing that only in combination can they create a truly scientific psychology. Description and correlational factor analysis are useful in delineating a taxonomy of personality, but they cannot settle arguments between rival taxonomies, and they lack theoretical support and experimental verification of such theories (H. J. Eysenck, 1984, 1985). If we are to establish a paradigm of personality research, we must proceed beyond taxonomy and factor analysis to the elaboration of *causal* theories and then experimental testing of such theories (H. J. Eysenck, 1983a). Such experiments may be along *behavioral* or *psychophysiological* lines (H. J. Eysenck, 1967), here we will be concerned solely with the latter.

An important signpost to suggest that psychophysiological variables are important for an understanding of personality is the well-established fact that genetic factors account for over half the phenotypic variance in most personality traits and types (Eaves, Eysenck, & Martin, 1989). But clearly DNA cannot influence behavior directly; it can only do so by exerting an influence on neuroanatomical structures, physiological processes, enzyme and hormone production, and similar biological phenomena. We clearly cannot rest content, with Skinner, in regarding neuroanatomy and psychophysiology as "black boxes" about which we know nothing other than that we feed stimuli into them and receive responses from them. This kind of psychology was already obsolescent in the 1950s, and is quite dead now; we do know something about cortical functioning, the autonomic system, the effects of hormones and enzymes, and the way in which these effects are integrated. We may not know as much as we would like to know, but to reject what we do know because it is short of complete knowledge, or in obedience to shibboleths from a distant past, would be foolish. We can only improve our knowledge by acknowledging its gaps and imperfections and strive to improve our ignorance; disregarding disdainfully what has already been accomplished, however small and poor the achievement, is not the way to scientific progress.

What we are aiming at, then, is a *nomological network* that spans the whole course from DNA through personality concepts to social behavior. I have elsewhere (H. J. Eysenck, 1993) given evidence for such a chain stretching from DNA through P to creativity and genius. Here I will suggest another such chain, grossly simplified, from DNA to sexual behavior (H. J. Eysenck, 1976). Genetic differences produce in some individuals an ascending reticular-activating system (ARAS) that overreacts or underreacts to neural input from collaterals of the ascending afferent pathways. The major function of the ARAS is to put the brain in a state of heightened arousal; if it is underreacting the cortex will be underaroused, if overreacting, the cortex will be overaroused (H. J. Eysenck, 1967). The picture is complicated by the fact that the ARAS receives input back from the cortex, informing it (e.g., that the information received via the ascending afferent pathways is repetitive and does not require high states of arousal). Equally, strong stimulation of the visceral brain or limbic system, which controls the expression of emotions and feelings, spills over into the ARAS, producing high cortical arousal (Figure 6). Theory states that extra-

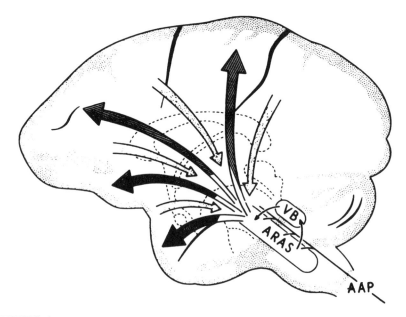

FIGURE 6

Visceral brain (VB) and ascending reticular-activating system (ARAS) as determinants of neuroticism and extraversion. AAP, ascending afferent pathway. (From H. J. Eysenck & M. W. Eysenck, 1985.)

verted behavior patterns are produced by a poorly functioning ARAS, leading to poor arousal potential; in other words, extraverts require more and stronger stimulation to attain a satisfactory level of cortical arousal. Conversely, introverts have an overactive ARAS, leading to a high arousal potential; in other words, introverts have to avoid strong stimulation that would create painful overarousal. Ambiverts, on this account would be intermediate.

Consequences of this theoretical account are shown in Figure 7. The abscissa shows level of stimulation, from low (sensory deprivation) to high (pain). The ordinate shows the hedonic tone (pleasant or unpleasant) produced by the arousal level associated with given degrees of stimulation; overall the regression is clearly curvilinear (i.e., extremes of low and high arousal are avoided). This much is widely agreed upon; as Berlyne (1960) points out, some such theory goes back to Wundt (1874), although without benefit of physiological measurements. What the theory states is that the curves of introverts will be tilted to the left (because of their high-arousal potential), and that of extraverts to the right (because of their low-arousal potential). Thus introverts will be predicted to tolerate sensory deprivation better than extraverts, while extraverts will tolerate pain better than introverts. Both predictions have been verified (Eysenck & Eysenck, 1985).

We can deduce from the theory that extraverts would look for excitement in the sexual field, to increase their arousal; such excitement would be provided by having sex early, more frequently, in more different positions, with more partners, and changing partners more quickly. All these predictions have

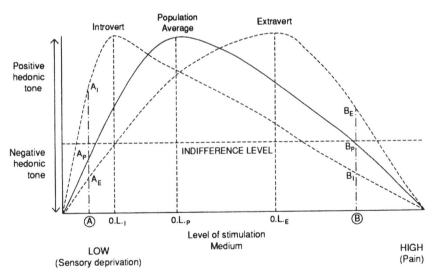

FIGURE 7

Hedonic tone as a curvilinear function of level of stimulation. I, introvert; E, extrovert, P. (From H. J. Eysenck and M. W. Eysenck, 1985.)

been verified (H. J. Eysenck, 1976), and thus we would predict that extraverts would be more likely than introverts to end up in the divorce court. This is a typical theoretical chain leading from DNA to social behavior; many others have been constructed and supported by empirical research (Wilson, 1981).

II. METHODOLOGY AND PROBLEMATICS OF PSYCHOPHYSIOLOGICAL RESEARCH

At first sight it must seem fairly easy to test a theory like that stated in the Introduction, or others to be mentioned later. However, there are so many problems and pitfalls that may not be apparent at first sight that a whole section will be devoted to their consideration. The first problem is the familiar one of "fuzzy theories." The concept of arousal at first sight may seem fairly clear, but research soon shows that its measurement and definition are anything but clear. The simple notion of collecting a group of males and females, measuring their degree of E by ratings, self-ratings, observed behavior or miniature situation tests, then taking them one at a time to the psychophysiological laboratory and measuring their arousal, and finally calculating the correlation between the two, encounters at least a dozen problems and difficulties, the major one being that there is little agreement on the best way to measure cortical arousal, quite apart from the fact that there may be several different types of arousal (Strelau & Eysenck, 1987). This is an important point to which we shall return later; it is not impossible to use "fuzzy concepts" in scientific research, but it does create very real problems.

Having noted this fundamental problem, which of course always appears when research in a given field is just beginning, and is not particular to psychology, we may note our second problem. This may be called the complexity of the organism. We never measure any dimension of personality in isolation; we are always dealing with an organism that differs from others not only in E (which is the dimension in which we are interested), but also in N, P, intelligence, as well as thousands of other variables. We may make a prediction on a person's sexual behavior based on his or her degree of E, but if the person is highly N, or very stable, sexual behavior may be sidetracked; it has been shown that N too is related along predictable lines with sexual behavior. So is P; high scorers have much higher libido than low scorers. Even octant analysis (looking at all possible combinations of P, E, and N) might not be enough; one person may be brought up in a very restrictive, religious atmosphere, another in a very liberal, permissive household, even though they fit into the same octant in terms of personality. Thus ideally we should measure *all* relevant variables and *all* their intercorrelations, linear as well as curvilinear, but for reasons of time and expense we are usually restricted to a small sample of all relevant variables—even if we did already know which variables are relevant!

Given that a particular behavior (and I shall include psychophysiological and other biological measurements under the term behavior) may be related to several different personality variables, we cannot expect that the variance *shared* between personality and behavior will be very large; in other words, we might expect correlations between extraversion and a measure of arousal (say low electroencephalogram (EEG) amplitude and high frequency of alpha) to be fairly low—anything above .4 would be unexpectedly high. But to obtain a value of .40 at $p < .01$ we need 40 subjects, and to narrow the fiducial limits of our correlations we would probably not be content with less than 100. If the true correlation is .30 (not an unlikely value), we would need 70 subjects to reach a $p < .01$ value! For a correlation of .20 (not unusual) we would need 150 subjects. Simple power considerations (Cohen, 1988, 1990) suggest that any such studies should use a number of subjects sufficiently large to guarantee that the expected correlation is also statistically significant. If you expect (from theory or previous research) that the expected correlation is around .30, then to use 40 subjects guarantees that a nonsignificant result would be uninterpretable! Yet very few psychophysiological studies pay any heed to the requirements of power analysis, and most use too few subjects, making negative outcomes meaningless. Because a study failed to demonstrate the predicted outcome at a statistically significant level does not *disprove* the hypothesis if the power of the test was insufficient—the study just was not worth doing! Many promising theories have been "disproved" in this fashion although the theories might have been perfectly correct. A review of the literature must thus carefully avoid Type Two errors based on this simple yet almost universal statistical misconception. The null hypothesis should never be accepted when the power of the test is clearly insufficient. In the absence of help from the author of a report, the reader must perform his own calculations to satisfy himself on this point.

Another reason why correlations will tend to be low, apart from human complexity, is the possible contamination of the arousal measurement by external factors. Time of day when the measure is taken may be a powerful determinant; arousal waxes and then wanes from morning to night, as we shall see, and if subjects are not all tested at the same time, this factor may play a prominent part in the measurement. Consumption of alcohol prior to testing may lower arousal; smoking may raise it; eating before testing may also have an effect. So may an event antecedent to testing, such as having a quarrel with the wife, hearing some good or bad news, sitting an examination, and so on. Arousal is very easily modulated; just seeing a pretty girl, or a dirty man, may have a profound effect. As all these events are largely uncontrollable—you can ask your subjects not to drink coffee or tea, but will they obey instructions? Will they avert their eyes from the pretty girl or the dirty man? All these variables, and hundreds more, increase the error variance to frightening proportions and ensure that observed correlations will not be high.

In addition to all these problems, there are some that are inherent in the experimental situation itself. Consider as an example of this danger a test of the arousal hypothesis that has become well known. I predicted that introverts, with high cortical arousal, should form conditional responses more quickly than extraverts; this follows directly from Pavlovian principles (H. J. Eysenck, 1967); I also predicted that N would not normally be correlated with speed of conditioning. Kenneth Spence (1964), on the other hand, used the Hullian formula $P = D \times H$ (performance is the product of drive and existing habits) to predict that conditioning should be a function of anxiety (N) considered as a drive; he saw no reason to expect E to be correlated with conditioning. Both sides produced evidence to support their claims; my colleagues and I found conditioning to correlate with introversion, but not with N, whereas Spence and his colleagues found conditioning to correlate with anxiety (N), but not with introversion. Each side succeeded in replicating their original studies several times (H. J. Eysenck, 1981).

Others, like Kimble (1961), also failed to replicate Spence's results, and finally Kimble visited Spence's laboratory to try and find a reason for these widely differing results. The answer became obvious. Kimble, like myself, had taken care to lower the natural anxiety of subjects coming to a psychological laboratory, assuring them that they would not receive electric shocks, no painful stimulus would be involved, that there was no apparatus visible, and more. Spence, on the other hand, did exactly the opposite—no reassurances were given, the apparatus with its frightening appearance was out in the open, and subjects were quite anxious when the trials began. In Spence's setup individual differences in anxiety were encouraged by the way the experiment was done, and the high anxiety aroused in some high-N subjects would produce much higher levels of cortical arousal than would be produced by introversion; hence individual differences in extraversion–introversion were drowned out by the much more powerful influence of anxiety. In our laboratories there was a minimal degree of anxiety, not sufficient to influence the experiment, and the arousal differences contrasting extraverts and introverts had free play! Spence never included a detailed description of his setup, with its anxiety-producing properties, in his accounts, so that all the apparent contradictions remained mysterious and inexplicable.

The problem is clearly one of state or trait measurement. Over 2000 years ago Cicero in his Tusculan Disputations pointed out the difference; a person may have a dispositional trait to evince anxiety easily, but may not do so in a particular nonfrightening situation (*state*), whereas another may have a low *trait* score, but nevertheless show anxiety in a given strongly fear-producing situation (*state*). Spence's setup encouraged differences in trait anxiety to emerge because of the frightening setup that produced state anxiety in those predisposed to show it. In my setup state anxiety was not aroused, and hence differences in trait anxiety had no possibility of becoming apparent.

The same argument applies to the measurement of physiological arousal

(Caccioppo & Tassinary, 1990; Coles, Donchin, & Porges, 1986). Consider another set of experiments. I had argued that if we used the EEG measures originally used to define arousal—high arousal is indexed by low-amplitude, high-frequency alpha waves—then introverts should show this pattern, while extraverts should show high-amplitude, low-frequency alpha waves. Gale (1983) reviewed 33 studies containing a total of 38 experimental comparisons. Extraverts were less aroused than introverts in 22 comparisons, introverts were less aroused than extraverts in five comparisons. A comparison of 22 for and five against a theory might be quite acceptable, but a scientist would demand a 100% success. What went wrong in the reluctant five experiments?

Let us consider the trait–state effect. The conditions under which the experiments were done varied widely, producing *state* conditions that might be expected to call for differing degrees of arousal. In some studies the subject relaxed in a semisomnolent state with his eyes closed, whereas in others the subject sat upright and attempted to solve complex arithmetic problems. Both extremes, as Gale suggested, would be expected to have an effect on arousal. The low-arousal setup resembles sensory deprivation, which is intolerable for extraverts (too little stimulation) and thus produces a high degree of arousal based on his discomfort with the situation; introverts would be more tolerant of such conditions, and less state-aroused. Hence in this situation extraverts might show higher arousal than introverts.

Next consider the most arousing conditions. Here the well-known Pavlovian law of *transmarginal inhibition* might be invoked; this law, also sometimes referred to as one of *protective inhibition,* states that an increase in stimulation is followed by an increase in reactivity *until* an optimal point is reached; beyond that inhibition sets in, and reactivity decreases. As we shall see, there is good evidence for such a law, and it has the corollary that this optimal point would occur at a lower level of stimulation for introverts than for extraverts, as will be obvious from a consideration of Figure 7. In the case of the most arousing testing conditions we might suspect that the optimal point for introverts, but not for extraverts, had already been passed, so that extraverts' arousal was still growing, introverts' declining.

Gale (1983), who advanced this argument, classified all the relevant EEG studies according to whether the conditions were minimally, moderately, or highly arousing. Introverts appeared to be more aroused than extraverts in all eight of the studies using moderately arousing conditions that reported significant effects of E, but this result was found in only nine out of twelve significant studies using low-arousal conditions and five out of seven using high-arousal conditions. Thus the five reversals of prediction occur in inappropriate conditions producing *state* conditions of arousal that may be opposed to trait predictions. Whether this is a true interpretation of the facts only an experiment actually *varying* conditions of testing can tell; Gale's theory would predict an inverse-U-shaped regression on EEG arousal–

introversion correlations as we go from low-arousing through intermediate to high-arousing conditions.

We may put the problem in another way. We are never testing introverts and extraverts but organisms whose reactions to experimental situations are *in part* determined by arousability. Consider a study by Tranel (1962) who tested my prediction that extraverts would be less able than introverts to tolerate sensory deprivation. In actual fact extraverts spent *more* time in the experimental room, thus apparently disproving the hypothesis. However, a more detailed analysis of the subjects' behavior under sensory-deprivation conditions pointed to a rather different conclusion. Subjects had been in-structed to lie quietly on their couches and not to fall asleep, but the extraverts tended to disregard these instructions. Some fell asleep, others moved about; they thus *increased* arousal level, driven by the predicted painful lowering of arousal in the sensory-deprivation condition. The *organism* reacts to change experimental conditions and thus makes the testing on introversion–extra-version interaction with the environment more difficult. In the EEG studies under minimum arousal conditions extraverts might react by producing vivid pornographic imagery to increase their arousal level; such activities are difficult to control; overt disobedience to the experimenter's rules, as in the Tranel experiment, may not be necessary to defeat his implicit intentions.

These are some of the problems of studying arousal effects in extraverts and introverts. Other problems arise in connection with the study of N. The main one is the difficulty of providing appropriate stimuli that would produce strong anxiety, depression, or worry in the laboratory. It would obviously be inadmissible ethically to induce strong emotions of this kind in our subjects, and in addition, how would we achieve this? We are thus restricted to ex-tremely mild manipulations of emotional variables, and they may simply be too weak to measure correlations with physiological effects that would be only too apparent were strong emotions involved (Lader, 1973; Lader & Wing, 1966). Even worse is the fact that different stimuli may produce contradictory effects. Thus Saltz (1970) showed that failure or the threat of failure, which is frequently used to produce anxiety in the laboratory, pro-duced more anxiety among high-anxiety subjects than among low-anxiety subjects, whereas electric shock, which is also frequently used to produce anxiety in the laboratory, generated more anxiety in low-anxiety than in high-anxiety subjects! Our correlations between N and psychophysiological effects may depend on the nature of the stimulus, and may reverse direction as we go from failure to shock. There is an absence of strong theories to explain such reversals and predict the direction expected in our experimental studies.

All these are problems on the personality–stimulus side. We have even worse problems on the psychophysiological side. Perhaps the major problem here is "response specificity" (Lacey, 1950, 1967; Lacey & Lacey, 1958); in responding to a given stressor, different physiological systems may be primarily activated. Some people may react to emotional stimuli primarily through an increase in heart rate, others through an increase in the electrical conductivity of the skin, others yet through more rapid breathing, or through the tensing of the musculature, and so forth. No single measure is adequate to portray the complexity of reactions; the recommended solution is to take measures of as many systems as possible, and index changes in the system maximally involved. Few experimenters have followed this advice, so that failure to support the theory may be due to faulty or too restricted choice of measuring instrument. (I have already mentioned "stimulus specificity" in connection with the work of Saltz, 1970; it should be noted that there is evidence for the genetic determination of *specific* fears and phobias, linked with restricted classes of feared objects or situations—H. J. Eysenck, 1987.)

The work of Thayer (1989) is exemplary in clarifying these issues. He contrasts two kinds of arousal, *energetic,* which is similar in many ways to cortical, introverted arousal, and *tense* arousal, which is similar in many ways to autonomic, neurotic arousal. In one study Thayer (1970) measured changes in heart rate (HR), finger blood volume, skin conductance, and muscle action potential, as well as obtaining self-ratings of arousal. Resting conditions were contrasted with arousing task conditions where subjects were required to count backwards from 100 by sixes or sevens while a loud buzzer was sounded, and while they were urged to count faster. Physiological changes from rest to task conditions occurred in between 62% and 93% of the cases for the four measures, as did self-ratings. Specificity was shown by the fact that the intercorrelations between the four measures of psychophysiological responding were very low; however, when physiological measures were combined to form a single general arousal index, the correlation of this index with the self-report measure was substantial. Thayer argues that the fact that self-report measures correlate substantially higher with combinations of physiological measures than the physiological measures correlated among themselves is of considerable theoretical interest. It not only suggests that general bodily arousal must be evaluated with multiple physiological systems, but it also suggests that self-report may be a better index of arousal than any single physiological measure. The potential superiority of self-report is further indicated because the most sensitive of the self-report measures recorded the expected change in 100% of the cases, and the best of the physiological measures (HR) showed change in 95% of the cases (and other physiological measures ranged to only 62% change). From this it would seem to follow that our usual mistrust of verbal (introspective) reports may be mistaken, the cortex seems able to integrate incoming messages from physiological systems quite well, and produce a veridical report about states of arousal more reliable and valid than any physiological channel can provide. (It should of course be remembered that Thayer did not use any cortical [EEG] measures; they might supply information superior to the measures actually used.)

Another problem in psychophysiological research is the existence of threshold and ceiling effects. If we take the electrodermal response as a measure of arousal, we may use (1) size of response, (2) latency of response, or (3) duration of response (i.e., time to return to baseline). The last-named, in the form of Freeman's (1948) recovery quotient (RQ), has been the most successful; he defines it in terms of:

$$RQ = \frac{B-C}{B-A} \tag{1}$$

where B is arbitrarily defined as the point reached by electrodermal response (EDR) 30 s after stimulation; point C is the level reached 5 min after peak mobilization (B), and A is the level at which the stimulus is applied. With time relations of A, B, and C in this constant ratio, the degree of homeostatic recovery is reliably indicated by dividing the percent-discharge decrement by the percent-mobilization increment that occurred in the standard periods of measurement. The reason why measures (A) and (B) may not be very valid is that high-N subjects may have a starting level well above that of low-N subjects, so that amount of change is restricted by ceiling effects. Latency of response too may be influenced by such factors; high-cortical arousal may lead to quick responses *ab initio* which can only be made faster to a limited degree. Statistical manipulation may be able to sort out these problems, but it makes assumptions that may be doubtful.

These problems should illustrate the difficulties of establishing the truth of theories such as those considered in this chapter. We are dealing with *weak* theories (H. J. Eysenck, 1960), as opposed to the *strong* theories often found in physics and other hard sciences. As Cohen and Nagel (1936) pointed out, in order to deduce the proposition (P) from our hypothesis (H), and in order to be able to test P experimentally, many *other* assumptions (K) must be made (about surrounding circumstances, instruments used, mediating concepts, etc.). Consequently it is never H alone that is being put to the test by the experiment—it is H and K together. In a weak theory, we cannot make the assumption that K is true, and consequently failure of the experiment may be due to wrong assumptions concerning K, rather than to H's being false (H. J. Eysenck, 1985). In a strong theory, enough is known about K to render empirical failure more threatening for H. This argument suggests that positive outcomes of testing H are more meaningful than negative ones. The former imply that both H and K are true; the latter that *either* H or K or both are false. In other words, failure does not lead to any certain conclusion. These considerations should be borne in mind when evaluating the material presented here. Even strong theories, such as Newton's theory of gravitation, were full of anomalies that occupied experimenters for 300 yr in an attempt to strengthen K. Psychology also needs a similar concentrated effort, involving what Kuhn (1970) calls "the ordinary business of science," in order to resolve the many anomalies still remaining.

In other words, if our theories are borne out by psychophysiological or behavioral and experimental tests, we may regard this as strong support (unless there are alternative theories making the same predictions). Failure to confirm may be due to any of the many possible causes I have listed, and should not be used to dismiss the theory. In view of the many candidates for the Nagel and Cohen "K," it might be thought that positive outcomes of such tasks would be extremely rare, but as we shall see this is not so—there are many positive results, suggesting that we are dealing with what Lakatos (Lakatos & Musgrave, 1970), calls a *progressive* problem shift as opposed to a *degenerative* problem shift, like the Freudian. The latter denotes a program of research that fights a rear-guard action by ad hoc explanations of anomalies that are destructive to the program.

I have discussed these philosophical considerations because psychologists are often naive concerning progress in science, assuming that theories are either right or wrong, and that deductions from such theories can easily decide whether they are one thing or the other. When success is not immediate, the theory is thrown away, and some other theory taken up for a while, until that too is rejected on possibly irrelevant grounds. Any theory as stated at the beginning is almost certainly wrong in important ways, but it is a good theory if it leads to important discoveries that will lead to its improvement. Dalton was the father of modern chemistry through his theory of atomic structure, but all that Dalton said about atoms—apart from the basic fact of their existence, which was not novel—was wrong; they are not indivisible nor of unique weight, they need not obey the law of definite or multiple proportions, and in any case his values for relative atomic weights of molecular contributions were for the most part incorrect (Greenaway, 1966). But these points were easily clarified in later research. The point is that he established a progressive problem shift which set modern chemistry on its feet. The nature of science is to correct early errors of thinking through improved research, not through throwing out theories that are less than perfect.

III. THE PSYCHOPHYSIOLOGY OF EXTRAVERSION–INTROVERSION

There are many reviews of the literature that go into considerable detail concerning general issues (Birbaumer & Schmidt, 1991) and specific details (Anderson, 1990; Eysenck, 1990; Gale & Eysenck, 1992; Neiss, 1988; Strelau & Eysenck, 1987; Stelmack, 1981; Strelau, Farley, & Gale, 1985; Zuckerman, 1991); these can be consulted with respect to earlier research that will only be summarized here. In general, the literature can be reviewed *either* by devoting separate sections to the various dimensions of personality and recording the evidence concerning their relation with different psychophysiological variables, *or* by devoting separate sections to the various psychophysiological parameters and recording the evidence concerning their relation with the different dimensions of personality. Neither approach is entirely satisfactory as the same experiment may involve more than one dimension of personality and more than one psychological measure. I have chosen the first alternative, as being psychologically more relevant, but where necessary I have also devoted sections to specific measures (e.g., differential effect of drugs).

Before starting out on a discussion of E, it is necessary to clarify the predictions involved. The theory comes in a strong and a weak form. The strong form has been stated already; the overly strong action of the ARAS in response to stimuli causes high states of cortical arousal, and hence introverted behavior, whereas the inhibited action of the ARAS in response to stimuli causes low cortical arousal, and hence extravert behavior. The weak form would simply link cortical arousal with introversion–extraversion differences, leaving open the causation of their differences in arousal level. It is difficult (but may not be impossible) to decide experimentally about the causal involvement of the ARAS; this is an important theoretical decision.

Linked to some extent with it is the question of whether E is linked to *tonic* or *phasic* levels of physiological activity. The tonic- or base-level measures refer to a general physiological state, usually a *resting level*, while phasic-response measures index the more or less transient response to stimulation. These measures are not always independent of each other, as I have already pointed out; neurotics may be in a permanently high state of physiological excitement (autonomic activity), which restricts possibilities of strong responsiveness to stimuli because of ceiling effects. Even during apparent rest, stimuli impinge on the subject from outside, and internal stimuli (imagination, worry) cannot be controlled by the experimenter. Does our theory refer to tonic or phasic levels of activity? Figure 6 should make it clear that the theory refers to phasic levels; it is messages traveling along the ascending afferent pathways that enter the ARAS via collaterals and set the arousal system in motion. Hence it may be more useful to talk about the *arousability* of the system, rather than arousal itself, which is of course largely a function of the strength of the stimulus. The effect of degree of modulation itself, of course, is given by Pavlov's law of transmarginal inhibition, already mentioned; this is also known as the Yerkes-Dodson Law (Broadhurst, 1959) or the inverted-U law. All this leads to the conclusion that we are dealing with the measurement of *phasic* arousal, that *tonic* arousal may also be involved, and that the effect of increasing stimulation will show itself in a curvilinear effect with extraverts showing the changeover at higher intensities than introverts.

The transmarginal inhibition effect is well authenticated, but also presents obvious difficulties; it is difficult, and may be impossible, to predict *at what point* in the stimulation continuum the turning point will come. Ideally, experiments should include the whole range of stimulation, from weak through medium to strong, including the reversal points for introverts and extraverts. An example of how experimenters should proceed (but usually don't!) is given by Shigehisa and Symons (1973) and Shigehisa, Shigehisa, and

Symons (1973). They measured auditory thresholds under 10 different intensities of illumination, the assumption being that (1) increasing light intensities would increase arousal, and hence lower sensory thresholds; that (2) transmarginal inhibition would reverse this trend at some level of intensity; and that (3) this point of reversal would occur earlier for introverts than extraverts. Results are shown in diagrammatic form in Figure 8; they confirm all the predictions. (In a second experiment conditions were reversed, i.e., auditory loudness was varied, and visual thresholds measured; predictions and results were identical.)

The most obvious psychophysiological test of the arousal theory has already been mentioned; EEG alpha is the standard measure of cortical arousal, ever since the discovery of the ARAS, and results, it will be recalled, tend to support the theory, especially when care is taken to use testing conditions neither too arousing nor too sleep making. The results illustrate the absolute necessity to control testing conditions in order to obtain meaningful results; unless this is done, the theory is not being tested.

A direct test of Gale's hypothesis has been reported by O'Gorman (O'Gorman, 1984; O'Gorman and Mallise, 1984; O'Gorman and Lloyd, 1987), using testing with eyes open and eyes closed as the testing condition. The earlier study obtained positive results, but has been criticized by Gale (1984), and the latter gave correlations only with impulsiveness, not with E; high impulsives are less aroused than low impulsives. The addition of covariates

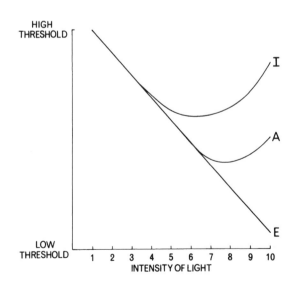

FIGURE 8

Auditory thresholds as a function of extraversion (E)–ambiversion (A)–introversion (I) and intensity of light. (From Shigehisa & Symons, 1973, with permission.)

generally heightened the observed relationship. Clearly the optimal conditions for obtaining positive results have not yet been sufficiently clarified. Another point to be considered in future work is the possibility that P may also be partly determined by cortical arousal (negatively), a possibility suggested by some of the studies to be considered later. It is also relevant that impulsivity correlates significantly more highly with P than with E.

Zuckerman (1991) gives a thorough review of the literature on EEG–E relations, pointing out the differences in judgment between Gale and O'Gorman as to what studies support and do not support the arousal hypothesis; he also points out that studies using only or mainly men seemed less likely to give positive results than studies using only or mainly women. While on the whole results are positive, and while many negative results can be explained in terms of Gale's hypothesis, we are not yet in a position to prescribe conditions that will always lead to positive results.

A paper by Matthews and Amelang (1992) looked specifically at the relation between EEG arousal and personality, also incorporating a variety of performance tasks. They concluded that the relationships between personality, arousal, and performance assumed by arousal theory are small in magnitude, which may explain inconsistencies in previous findings. Interactive effects of E and cortical arousal on attentional processes appear to be rather more reliable. Possibly the most interesting finding in this study was the P was significantly related with low arousal; we have already noted several suggestions that this might be so.

EEG measures have an important advantage—they are very much determined by genetics. Bouchard (1991) summarizes literature to show identical intraclass correlations across the four classic EEG bands of .80 for monozygotic (MZ) twins brought up apart, and .81 for MZ twins brought up together. The latest results reported by Stassen, Lykken, and Bomben (1988) and Stassen, Lykken, Propping, and Bomben, (1988) show that an identical twin gives results as much like the other twin as he does with a second testing of himself! Hence disattenuated heritability estimates cannot be far removed from unity. Few psychological measures are so firmly grounded in heritability.

Direct EEG measurement has given results that are replicated with fair accuracy by averaged cortical evoked responses to auditory stimulation. Introverts exhibit increased auditory evoked potential (AEP) responses to tones of moderate intensity, such as low-frequency (500 Hz) stimuli of 80-dB intensity; greater individual response variability to lower frequency than to higher auditory frequency stimulation facilitates this effect (Stelmack, Achorn, & Michaud, 1977). It has equally been found that effects are more easily produced when auditory stimuli are varied within a series; this can be done either by alternating high and low frequency or by randomly varying the sequence of intensity levels (Brunean, Roux, Perse, & Lelord, 1984). Finally, when the interstimulus interval is short, the enhanced response of introverts occurs only with the first stimulus in a repetitive series (Stelmack, Michaud,

& Achorn, 1985), suggesting that neural response recovery is incomplete. All these results hang together in a meaningful and predictable pattern.

Work using electrodermal recording methods firmly supports EEG studies in that significant confirmations of the theory are usually obtained when auditory stimuli in the middle range (75–90 dB) are used. Under conditions of low-intensity auditory stimulation (60 dB or less) no significant differences between extraverts and introverts are typically found, whereas with higher intensity stimulation the expected transmarginal inhibition sets in, and skin conductance responses show greater amplitude in extraverts than in introverts. The differences in the medium intensity range are shown by introverts having initially greater response amplitude, slower habituation rate, more frequent number of responses, and greater response to test stimuli following a repetitive habituation series. Stelmack (1990) cites 15 separate studies to support these generalizations; not every study shows all these effects, but there is practically unanimity in the overall effects.

Of particular interest in relation to skin conductance and personality is a study by Wilson (1990) that was carried out, not in the laboratory but during the subjects' everyday activities, thus avoiding the usual accusation that laboratory conditions are artificial and tell us little about events of ordinary life experiences. It also avoids the problem of time-of-day effects by recording skin conductance every hour from 7 AM to midnight, by the use of a self-monitoring device activated once every hour. In addition, subjects recorded the activities they had been engaged in at the time, and listed any drugs they had taken that might affect recordings, such as coffee, alcohol, cigarettes. Figure 9 shows the most important results; the age-correlated introversion curve should be used to compare the two types, as the number of sweat glands in the skin declines with advancing age, even at the fairly young level of Wilson's subjects.

It will be clear that (1) throughout the day and until 9:00 PM, introverts show greater arousal than extraverts; (2) there is an equal time-of-day effect for both types, with a high between 3:00 and 4:00 PM; (3) the curves come together in the evening, and coalesce after about 10:00 PM. Records of activity show that this evening effect is due largely to the greater activity of extraverts in the evening (going out, discos, concerts, etc.) as compared with introverts (reading, TV, resting), itself presumably produced by the low-arousal person's need for increasing arousal. Conversely, introverts are more active in the morning, which may account for the large gap between the curves from 7:00 AM to 9:00 AM. During the day, of course, most people's activity is circumscribed by the demands of their job, and hence they are little able to influence this arousal level. This study thus adds valuable evidence to the laboratory studies already surveyed.

Stelmack (1990) summarizes this set of studies by saying that the work using electrodermal measurement techniques complements the work using EEG auditory evoked response measures; all these approaches "are in general

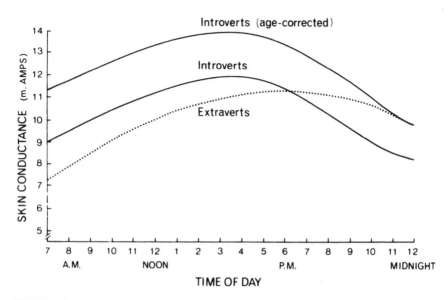

FIGURE 9

Skin conductance (arousal) level of introverts and extraverts during course of day. (From Wilson, 1990, with permission.)

agreement with Eysenck's hypothesis that introverts are characterized by higher levels of arousal than extraverts, at least when they are exposed to stimuli of moderate intensity or arousal value" (p. 298).

Another study using hourly testing, but for six separate days, has been reported by Thayer, Takahashi, and Pauli (1988). The Thayer measures for energetic arousal (E?) and tense arousal (N?) were used repeatedly. "An expected pattern of higher levels of energetic arousal for introverts than for extraverts did occur in a very general way, but the obtained differences were not statistically significant . . . and energetic arousal did not show the interaction effect with time of day that might be expected on the basis of earlier research by Blake (1967) using body temperature. However, tense arousal did show this effect" (p. 19). This study is interesting but the statistical power is too small to justify any conclusions, n being 21; a correlation would have to be .54 to be significant at the $p < .01$ level! The existence of a weak, negative relationship between E and morning–evening preference is supported by a report by Neubauer (1992), and seems firmly established by many reports summarized by Neubauer.

The argument concerning the use of moderate intensity stimuli may be extended to include another type of EEG recording, namely, the so-called contingent negative variation (CNV). This is a measure of electrocortical activity that is usually considered as an index of preparation for a motor response. In the typical paradigm a warning stimulus (S_1) is followed after an interval between 1 s and 4 s by an effective stimulus (S) which provokes a response (R); the CNV is the EEG response to S_1. This response, which develops in the interval between the warning and the imperative signal, is greater for extraverts than introverts (Lolas & Andrassa, 1977; Plorij-van Gorsel & Janssen, 1978; Werne, Favery, & Janssen, 1973). (This type of response will be taken up again when we turn to the consideration of the drug postulate in a later section.)

Several recent studies have reported interesting results concerning CNV correlations with personality. Thus Dincheva, Piperova-Dalbokova, and Kolev (1984) showed that introverts are less distractible, in terms of CNV changes, than extraverts; Werre (1983) reported that distraction smoothed the differences in CNV between extraverts and introverts. Piperova-Dalbokova, Dincheva, and Urgelles (1984) thought that this might be due to the greater effect of distraction on extraverts. They argued that the effect of distraction on CNV depended on the stability of the CNV: the more stable or strong the CNV "modeling system" (Sokolov, 1963), the more difficult it would be to effect changes in it. To test this assumption, they investigated the ability to pass from one task to another in extraverted and introverted subjects. The results supported the assumption that in introverts the CNV is more stable, so that the original modeling system is still active after the paradigm is changed. Thus a greater "changeability" of extraverts, already postulated by Wundt (1874), is also apparent in the electrophysiological phenomena of the CNV (Dincheva & Piperova-Dalbokova, 1982; Dincheva et al., 1984). (Wundt had argued that the four temperaments of the ancient Greeks could be conceived of as being produced by two orthogonal factors—the cholerics and the sanguines being "changeable," the melancholics and phlegmatics being "unchangeable." This corresponds to the E dimension. Cholerics and melancholics were emotional, sanguines and phlegmatics nonemotional; this corresponds to the N dimension.)

The latest summary and report of research on the CNV comes in a chapter by Werre (1986). He found correlations with E, but not with N. The clearest differential effect on the CNV occurred when isolated, motivated young adult students performed a constant four-period reaction time (RT) task, which was novel to them, in the morning. Under those conditions, there was a positive correlation between CNV amplitude and E. Conditions not producing personality differences were those that existed when the students were repeating the standard task, and when they were performing a second task in addition to the standard one. This study is notable for varying conditions under which subjects performed, so that the author could clarify the relationship between conditions and personality in effects on the CNV. (See also Werre, 1983.)

These CNV results suggest that differences between extraverts and introverts may also be found in the mechanisms that serve motor preparation. Stelmack (1985b) has contributed an extensive review of this topic, and the main results may be illustrated by reference to an examination of individual differences in spinal motorneuronal excitability (Pivik, Stelmack, & Blysma, 1988). These workers recorded the response characteristics of component muscle action potentials from the calf muscle of the leg, using electromyographic recording methods. Motor reflex responses are elicited by brief electrical impulses, and the time intervals between pulses are varied to obtain an index of reflex recovery. Extraverts showed less motorneuronal recovery (or excitability) than did introverts.

These results indicate that differences in E are related to discrete levels of CNS motor activity; this decreased motor neuronal excitability is associated with increased dopaminergic activity (Goode & Manning, 1985), a finding in good account with evidence linking E to low levels of monoamine oxidase and dopamine-beta-hydroxylase (DβH) in the noradrenergic system (Schalling, Edman, Asberg, & Oreland, 1988), neurochemical variations that can be associated with increased dopaminergic activity. As Stelmack (1990) points out, "these effects provide some plausible evidence that Extraversion may be characterized by fundamental differences in the expression of motor activity that may be referred to as discrete levels of motorneuronal excitability" (p. 308).

The results of Pivik, Stelmack, & Blysma (1988), have been replicated by Britt and Blumenthal (1991), who examined differences between extraverts and introverts in motorneural sensitivity to gradations in stimulus intensity. Startle stimuli (noise burst at 85 dB and 60 dB were presented with and without a prepulse. Startle response latency revealed that introverts responded more quickly to the 85-dB stimuli than the 60-dB ones, whereas extraverts showed no differences in response to the two classes of stimuli. Response amplitude or probability was not affected by personality; it is argued that the results support the view that "extraverts may possess an insensitive motorneuronal system" (p. 391).

A particular type of event-related potential (ERP) that has come to the front in recent years is the P300, which is considered to be a manifestation of cognitive activity because it is generated when subjects are required to attend and discriminate stimulus events differing from one another on some dimension (Polich, 1986). Such discriminations produce a large (sa. $10-20/\mu V$) positive-going potential with a latency of approximately 300 ms elicited with simple auditory stimuli in young adults. It is believed to reflect those processes that are involved when the mental model (or schema) of the stimulus environment is refreshed and updated (Donchin & Coles, 1988). This view relates back to Sokolov's (1963) model of the startle response and habituation, suggesting that the P300 is produced whenever attentional resources are needed to process new stimuli (Polich, 1989). P300 latency measures are usually interpreted in the sense that the peak of the P300 component is a measure of the *speed* of stimulus classification resulting from the discrimination of one event from another (Magliero, Bashore, Coles, & Donchin, 1989), and hence not related directly to behavioral RT measures. The fact that the activities associated with the P300 (habituation, orienting responses, stimulus classification) are closely related to the concept of cortical arousal suggests that tests of the introversion–arousal hypothesis could use the P300 paradigm with advantage.

The first directly relevant study in this connection was contributed by Daruna, Karrer, and Rosen (1985), who compared P300 ERPs from introverts and extraverts using an auditory stimulus guessing task with a large number of trials, finding that P300 amplitude was significantly larger for introverts. They interpreted this finding to mean that introverts attended to the task more than extraverts, even though task performance did not discriminate between the groups. The stress of the interpretation was thus on the allocation of attentional resources, suggesting that introverts concentrate that much more on the given task.

Pritchard (1989), using an auditory oddball task, failed to replicate the finding of an inverse relation between P300 amplitude of extraversion; P300 latency was directly related to extraversion in males but may have been secondary to a negative correlation in the sample used between E and N. P300 latency was inversely related to N, as had been found previously by Plorij-van Gorsel (1981), using a varied RT paradigm. O'Connor (1983), also using a varied RT paradigm, found the P300 larger in introverts than extraverts. Differences in subject selection and stimulus choice make these studies difficult to compare.

More clearly focused on a specific theoretical issue was a recent study by Di Traglia and Polich (1991), directed at Eysenck's suggestion that *habituation* would be faster in extraverts, when AEPs were elicited using a simple auditory discrimination task. Daruna et al. (1985) had already found that introverts showed greater amplitudes in their responses. Di Traglia and Polich found that introverts demonstrated greater amplitudes than extraverts *only after subjects had spent sufficient time on the tasks to produce habituation.* Time on task interacted with the personality group since P300 amplitude for introverts maintained the same amplitude over trial blocks, whereas extraverts showed a decline over trial blocks. This finding links the experimental results squarely with the greater *vigilance* usually shown by introverts (H. J. Eysenck & M. W. Eysenck, 1985). A review of the literature covering the AEP and vigilance phenomena is given by Koelega and Verbaten (1991).

Another more recent study of the P300 (Polich & Martin, 1992) gave some support to previous findings, in that male subjects showed greater P300 amplitude for introverts than extraverts; females failed to show this trend. Correlations for N and P300 latency went in the opposite direction to the trends observed in the Pritchard study. It is possible that had Polich and Martin used a more complex task than a simple target tone detection, or had used the vigilance paradigm, different results might have been obtained. The P300 is probably a valuable tool for personality theory testing, but experiments clearly require to be set up with more clear-cut theoretical guidance.

The latest study in this series (Ortiz and Maojo, 1993) used extreme scorers on extraversion–introversion, engaged in auditory oddball tasks. Latency and amplitude of P300 wave were recorded and data analyzed using nonparametric techniques. "Significant differences were found in the electrodes at Fz, Cz, and Pz sites, with larger amplitude for introverts than for extraverts" (p. 110). No significant differences were found for P300 latencies, confirming earlier work (O'Connor, 1983; Daruna et al., 1985), but contrary to Pritchard (1989). The observed differences in amplitude are quite large, amounting to over three times the size of the SD.

As an example of what should be done is the work of a group at Trier University. A particularly interesting use of the P300 paradigm has been reported by Bartussek, Naumann, Collet, and Moeller (1990). Essentially, they designed their experiment to test Gray's (1981) hypothesis that the behavioral activation system (BAS) is more active in extraverts, while the behavioral inhibition system (BIS) is more active in introverts, harking back to Eysenck's (1957) theory. This leads to the prediction that extraverts react more strongly to reward, introverts to punishment. In the experiment, AEPs were recorded to neutral stimuli (tones), as well as to tones signifying gains or losses of 0.50 or 5.00 marks on a betting task. Amplitude of the P300 was greater in introverts when neutral stimuli were administered; when the stimuli signified gain or loss of money extraverts showed greater amplitude, with the greater gains or losses producing greater effects than the smaller sums. The fact that gains and losses had identical effects contradicts Gray's theory. The findings agree with Eysenck's theory, if we assume that the arousal produced by the monetary gains and losses pushes introverts beyond the point when transmarginal inhibition sets in; the greater amplitude of P300 reactions for introverts encountering the neutral tone is also in line with the prediction. High-N scorers show greater amplitude of P300 than low-N scorers to tones signifying monetary returns (either gains or losses), but no difference for neutral tones.

An experiment planned along similar lines has been reported by Bartussek, Naumann, Moeller, Vogelbacher, and Diedrich (1990). Here stimuli were adjectives with emotionally positive, neutral, or negative meaning; instructions were either *cognitive* (word is longer, shorter, or equal to six letters) or *emotional* (word is emotionally positive, neutral, or negative). Regardless of instructions, emotionally positive and negative words produced greater P300 amplitude than did neutral words. Results for personality differences again disproved the Gray hypothesis; extraverts showed a stronger reaction to *both* positive and negative words than did introverts. The authors suggest that results are more in line with the simple arousal theory. For neutral words introverts show greater amplitude, as expected. These studies show interesting developments in the theory-related definition of stimulus conditions, a development that will undoubtedly improve our understanding of this important connection.

In an experimentally more satisfactory study, Bartussek, Diedrich, Naumann, and Collet (1993) repeated their attempt to test Gray's hypothesis. Reward and punishment were again operationalized by winning and losing different amounts of money in a gambling situation while measuring ERP to the stimuli signaling winnings and losses, and to the presentation of the amount involved. A strong interaction was found between E and winning/losing for the P2 amplitude of the ERPs to the tones providing feedback about winning or losing, just as predicted by theory. As they say, "if one accepts that more positive amplitudes in this time region are caused by a higher emotional significance of the stimuli (as suggested by the enhanced-P2 amplitude when winning or losing a higher amount of money), then for extraverts the feedback of winning has a higher emotional significance than the feedback of losing, while for introverts the opposite holds true—just as predicted by Gray's theory" (p. 573).

This experiment illustrates some of the characteristics of a good, as opposed to a poor experiment in psychophysiology. In the first place, there is a *specific* hypothesis, rather than a very general statement. In the second place, this hypothesis determines the experimental situation. In the third place, personality (extraversion or introversion) is defined by extreme scorers in either direction, thus enhancing the probability of significant results (statistical power). In the fourth place, the experiment contains an external variable (amount won or lost) that helps to define the meaning of the observed waveform. And finally, the experiment is not just a "one-off," but presents the last of a whole series of studies, improving and clarifying the situation with each step. If only all the experiments here described had been done with similar care and regard to theoretical and experimental requirements, our knowledge would be much further advanced!

These reports receive some support from a study by Pascalis and Montirosso (1988), who found that extraverts showed higher evoked potentials (EP) in a meaningful-speech condition, introverts in a meaningless-speech condition. Self-ratings of involvements agreed with the AEP recordings with respect to personality differences.

Any euphoria caused by the data surveyed so far would almost certainly be doused by considering the results of experiments using brain stem-evoked responses (BERs) (i.e., responses developing within the first 10 ms of stimulation). Their genesis is better understood than that of the usual waveforms developing within 100 ms to 500 ms of stimulation. The first two waves of the BER can be traced back to the auditory nerve and the cochlear nucleus, respectively, whereas wave five is determined primarily from activity at the inferior colliculus. Characteristically these waves are highly invariant across different stages of arousal, attention, sleep, and even metabolic coma (Stelmack & Geen, 1992). Stelmack and Wilson (1982) compared introverts and extraverts in their auditory BER to brief click stimuli varying in intensity from 55–90 dB and found positive correlations between extraversion and wave/latency at the 75-, 80-, and 85-dB intensity levels, thus again indicating an association of introversion with greater sensory activity, an association also reported by Andress and Church (1981), Campbell, Barikean-Braun, and Braun (1981), and by Szelenberger (1983). These results are theoretically important because BER effects are quite independent of the ARAS and cannot therefore be predicted from the strong theory. As Stelmack (1990) points out, these BER effects may require the elaboration of the neurological basis of extraversion from the brain stem reticular formation and mechanism of the limbic system to accommodate differences in arousal or synaptic transmission that are present in peripheral nervous system processes (Stelmack, 1985a).

The studies so far referred to are probably the most directly relevant to the arousal theory of E, but there are many others that lend support to the theory. One of the most revealing is the stimulated-salivation paradigm (S. B. G. Eysenck and H. J. Eysenck, 1967a). Rate of salivary secretion is measured in a resting situation, and again after four drops of lemon juice are deposited on the tongue; the increment in salivation is measured and related to E. Theory dictates that introverts would show greater incrementation than extraverts, and many studies (reviewed by Deary, Ramsay, Wilson, & Riad, 1988) have shown this to be so. Figure 10 shows the results of the S. B. G. Eysenck and H. J. Eysenck (1967a) study, carried out on 50 males and 50 females, each group divided into 5 sets of 10 each, according to their E scores. A correlation of −.71 was obtained, supporting the hypothesis. Deary et al. (1988) found an even higher correlation of −.74, but only in the morning.

In order to test the transmarginal inhibition hypothesis, S. B. G. Eysenck and H. J. Eysenck (1967b) repeated the experiment with subjects having to swallow the drops of lemon juice in order to increase the strength of stimulation beyond the reversal point. The prediction that now extraverts would show greater incrementation was also borne out at a high level of statistical significance.

The study of augmenting–reducing effects (Buchsbaum & Haier, 1983) is another line of research providing evidence both for the general theory and for the importance of transmarginal inhibition. The technique used assesses cortical (EEG) responses to stimuli of varying intensity. "Augmenters" are so called because the stimulus–response slope is positive (i.e., increased stimulation produces increased amplitude), whereas for "reducers" the slope is negative, (i.e., increased stimulation produces decreased amplitude). An early study by Zuckerman, Murtaugh, and Siegel (1974) illustrates the observed results, using scores on the sensation-seeking "high disinhibition" and "low disinhibition" scales as approximation to E and I personality. As expected, extraverts are augmenters, continuing to increase EEG amplitude with increases in stimulus intensity, whereas introverts, after an initial rise, reduce the EEG amplitude with increasing stimulus intensity. Figure 11 shows the results.

An experiment by Haier, Robinson, Bradley, and Williams (1984) failed to

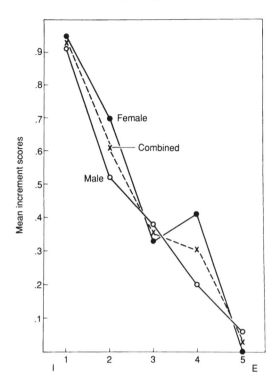

FIGURE 10

Increments in salivation of 50 male and 50 female subjects as a function of extraversion–introversion (E, I). (From S. B. J. Eysenck & H. J. Eysenck, 1967a.)

support this conclusion, but later work has tended to support the theory. Thus Lukas (1987) used a Maxwellian-View Optical System for precise control of retinal illumination, and found that while the occipital potential showed no correlation with the personality measures, the vertex potentials, which are known to be affected by nonsensory factors such as cortical arousal and attention, are very significantly correlated with the personality, in the sense that vertex augmenters are sensation seekers. This relationship, as expected, only appeared in connection with the more intense light flashes, very much as in the Zuckerman et al. (1974) study. He also concluded that how the brain responds to intense sensory stimulation determines how people respond behaviorally to intense sensations. Saxton, Seigal, and Lukas (1987) extended this work to cats and found that feline augmenters at a high-intensity range were likewise more active and exploratory, suggesting that the relationship between sensory modulation and behavior is not confined to humans.

The most complete study of the augmenting–arousing effects in relation to personality has been done by Stenberg, Rosen, and Risberg (1988). They used

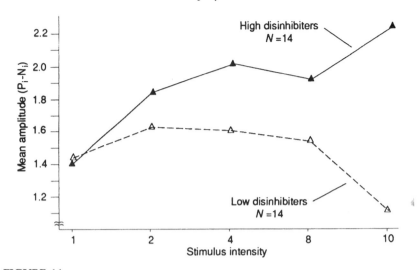

FIGURE 11

Mean average evoked potential (AEP) amplitude for low and high disinhibition scorers at various levels of stimulus intensity. Amplitudes are in arbitrary mm deflection units. Each mm unit = .42/μV. (From Zuckerman et al., 1974.)

six intensities of visual and six intensities of auditory stimuli, arguing that a superimposed mechanism of transmarginal or protective inhibition can account for the relationship with personality only if it generalizes across different modality and response definitions. They found that for the visual stimuli, the slope of the P90–N120 amplitude at the vertex correlated significantly with both the EPQ E Scale and the Disinhibition subscale of the Zuckerman Sensation-Seeking Scale in the way that augmenting–reducing theory would predict. However, over the primary visual area, no component showed the same personality relationship as the vertex wave, and one early component showed the opposite. This result, they argue, suggests that personality differences in visual EPs (VEPs) may reflect different ways of allocating processing resources between primary and association areas, rather than a general tendency to inhibit strong stimuli. In the auditory modality, personality differences were not apparent in the amplitude slopes, possibly due to the confluence from primary and association areas in AEPs in the vertex lead. The authors finally state that "there was a general tendency for latencies to correlate positively with extraversion and disinhibition, in congruence with Eysenck's theory on the biological basis of extraversion" (Stenberg et al., 1988, p. 571). These results powerfully indicate the need for very detailed and specific studies to look at the precise relationships involved, using different stimulus modalities and different recording areas; only by thus enlarging our understanding will we be able to arrive at a more all-embracing theoretical formulation than is available at present. Perhaps an explanation will be found along these lines for the curious and unusual results of the Haier et al. (1984) study.

Zuckerman, Simons, and Como (1988) have reported another recent study along these lines. They used 54 male subjects scoring high or low on the disinhibition scale; these were exposed to four intensities of auditory stimuli on one occasion, and visual stimuli on another. Interstimulus intervals of 17 s and 2 s were used. High disinhibitors (extraverts) showed EEG augmenting and low disinhibitors showed EEG reducing on three out of four series. AEP and VEP slope measures were correlated only for the long interstimulus interval (ISI) series. Clearly the theory is along the right lines, but we still do not know precisely which are the right parameters requiring control.

In a further study, Stenberg, Rosen, and Risberg (1990) extended their earlier study (1988), using differences in the direction of attention tò clarify the situation. As they stated, "The evoked response has normally been measured from association areas of the brain (at the vertex)" (p. 571). The present study measured VEP amplitude over visual cortex and at the vertex, using four light intensities in two conditions, where attention was either directed towards the light stimuli, or away from them by a concurrent auditory task. Forty subjects were classified as extraverts or introverts based on the Eysenck Personality Inventory. The results show that attention interacted significantly with E. Introverts exhibited a narrower focus of attention, with higher amplitudes and amplitude-intensity functions when attending to the light flashes and lower when distracted. Extraverts show smaller differences between conditions, indicating a more evenly distributed attention. Higher arousal in introverts is the probable cause of their narrower focus of attention. There were marked differences in the distribution of activity between vertex and occipital cortex. Introverts showed relatively stronger occipital responses and extraverts stronger vertex responses across all intensities and in both conditions. The predisposition for mainly perceptual responses to aversive stimuli in introverts, and for general alerting and motor-preparatory responses in extraverts, are interpreted as supportive of Brebner and Cooper's hypothesis that introverts are "geared to inspect" and extraverts "geared to respond." This study thus combines the large body of work suggesting that introverts tend to focus attention more narrowly (H. J. Eysenck & S. B. G. Eysenck, 1985) and also lends support to the Brebner and Cooper (1978) theory that excitation in extraverts and introverts is associated with motor responses and cognitive processes, respectively.

Response to photic driving has been a subject of considerable theoretical interest due to the work of Robinson (1982, 1986). In his work, visual stimuli were sinusoidally modulated, and the relative amplitudes of these sinusoidal EEG responses were measured for each stimulus frequency. Data were analyzed to produce measures of inductance (L) and capacitance (C), the C measure being regarded as an index of inhibition, the L measure as an index of excitation. Individuals in whom the excitatory and inhibitory processes were of comparable strength were regarded as balanced. Combining theoretical ideas from Pavlov and Eysenck, Robinson argued that stable extraverts are strongly balanced individuals, and neurotic introverts weakly balanced individuals. When only those subjects whose N and E score were comparable, and whose CP and L score were also similar, were compared, a remarkably high correlation of .85 between personality and the EEG measure was obtained. (Even including the whole sample still gave a correlation of .63.) Replication is of course needed before theory or correlation can be accepted as other than an intriguingly new approach.

Along more orthodox lines has been the work of Mangan, Pellett, Colrain, and Bates (1993), which has attempted to bring together ideas associated with Pavlov and Eysenck (Mangan, 1982). Mangan et al. investigated correlations between personality and driving in the different EEG bands, extracting four factors from a Principal Components analysis of correlations between driving power at the fourteen frequencies used. These factors were identifiable as (1) a middle-frequency factor embracing alpha and beta 1 bands; (2) a high-frequency factor, embracing most particularly beta 2 and gamma, (3) Factor 3 and 4 are theta factors, each with a loading on the adjacent (alpha) frequency of 9 Hz.

In relating personality dimensions to PD, P, E, and N scores were correlated with mean power scores in the five-frequency bands. Notable correlations were between P and gamma power, and between E and theta power: N failed to correlate with any PD factor. The loadings corresponding to these associations are very solid. On factor 2, P has a loading of .83; the gamma band of .83. Similarly, on factor 3, E has a loading of .73, the theta band of .88. It is clear that P is associated with high-frequency and E with low-frequency driving. What do these associations mean?

Mangan et al. suggest that the finding of orthogonal photic driving factors involving low (theta), middle (alpha and beta 1), and high (beta 2 and gamma) frequencies suggests that different structures are involved in photic driving in these frequency bands. A theory is suggested relating arousability in hierarchically organized brain systems to the hierarchy of personality dimensions E, N, and P, these traits reflecting strength in the brain stem, limbic, and cortical systems respectively. In accordance with this theory, they expect to find a relationship between these personality traits and photic driving at the "resonant" or natural frequency of these systems. For E and P the results seem confirmatory, but for N these seem to bear no relation at all. The close relation between these results and much Russian work, and the interpretation of Western and Eastern theories, make the study particularly interesting, but also particularly difficult to evaluate. Perhaps a more detailed statement of the theory, and a replication of the findings, will make the task easier.

A more direct and nontheoretical approach also using photic driving has been reported by Golding and Richards (1985). Their most convincing result was that P, rather than E, was related to low arousability; Deary et al. (1988)

had also found a significant correlation linking P with low arousability in their study of stimulated salivation already mentioned. Unfortunately the number of subjects in the Golding and Richards study was too small to allow great confidence in their findings.

A rather different approach from that using EEG measures involves regional cerebral blood flow (CBF), which has been shown to reflect differences in cortical activity related to mental effort and task execution (Ingwar & Risberg, 1965, 1967; Risberg, 1980, 1986). It is well established that the perfusion of the brain tissue is closely regulated by its functional demands (Raichle, Grubb, Gado, Eichling, & Ter-Pogonian, 1976). Indices of regional cerebral blood flow (rCBF) are obtained by tracing the clearance of an inert radioisotope from an array of detectors positioned around the head. This method was first used to investigate arousal differences between introverts and extraverts by Matthews, Weinman, and Barr (1984), using 51 women as subjects. In agreement with the arousal theory, they found negative correlations between extraversion and rCBF averaging −.36; the correlations were approximately uniform over all brain regions examined, eight for each hemisphere.

In a replication study, Stenberg, Risberg, Warkentin, and Rosen (1990) used 19 men and 18 women. Thirty-two detectors in all were used in the study. There was no overall correlation between E or N and blood flow, although by restricting the sample to women and partialling out age, Stenberg et al. obtained a correlation of −.34 with E, very much in line with Matthews et al. (1984). However, the main finding of the present study was in relation to regional differences in blood flow. Aggregating distribution values into five regions (frontal, temporal, central, parietal, and occipital), Stenberg et al. found a significant extraversion-by-regions interaction ($p < .002$), indicating different flow distribution between introverts and extraverts; these were replicated over hemispheres. The temporal lobes showed bilaterally higher activity in introverts, a result possibly explicable by their suggested role in sensory analysis, memory retrieval, and expressions of emotion, for which their closer connection with the limbic system, through the amygdalae and the hippocampi, provides an anatomical foundation. It is noteworthy that direct stimulation of the temporal lobes in nonhumans produces a state of behavioral inhibition, hypervigilance, and facial expression indicating arousal (Kaada, 1960).

It is too early to come to firm conclusions. Sex and age differences are clearly important, and need to be taken into account; women and older people may show an overall effect where men and younger people may not. Of particular interest are the regional differences, which agree well with predictions derived from the general arousal theory, and more particularly from the Brebner and Cooper (1978) extension relating E to motor functions and introversion to cognitive function. Future work dedicated to clarification along these lines should be of considerable interest.

Heart rate (HR) provides another possible index of cortical arousal. Pearson and Freeman (1991) tested 15 introverts and 15 extraverts, varying the level of difficulty on an arithmetic task. HR, the dependent variable, increased with task difficulty, and introverts showed larger HR reactivity to the task than extraverts. The difference arose because baseline levels for the two groups were identical, and extraverts showed no change in HR during the task, whereas introverts showed significantly higher HR as a consequence of performing the task. Differences were least under the easy condition and equally large under the moderate and difficult conditions. In earlier studies, Geen (1984) had found a curvilinear relationship between intensity of noise and HR response, introverts having higher HR in response to a noise of an intermediate intensity than extraverts. Glass et al. (1984) failed to find differences in HR activity in response to a series of subtraction tasks and a modified Stroop test, possibly because of the low statistical power of his design. Overall the results support Houston's (1986) general suggestion that introverts should show greater HR reactivity than extraverts as measured by HR but not by any measured HR variability.

A more recent test of the hypothesis by Roger and Jamieson (1988) failed to show significant correlations between E or N and HR responses to the Stroop test. More successful was a study by Richards and Eves (1991). In their experiment, biologically based personality measures were compared between subjects who showed secondary accelerative cardiac-defensive responding (accelerators) and subjects whose HR remained relatively unchanged (nonaccelerators) during the interval following high-intensity auditory stimulation. Introverts (low E), and subjects with high N scores showed higher levels of HR throughout the experiment than extraverts and stable subjects (see also Haier & Hirschmann, 1980).

These results agree well with expectation. Similar positive results were obtained by Gilbert (1991) who recorded HR during venipuncture (a passive task) and speech-making (an active coping task). She found that task was a major determiner of the relationship between HR and personality. During venipuncture HR increase was positively correlated with I and N. However, during the speech HR changes were negatively correlated with introversion. Skin conductance changes were weaker but, when occurring, were in the same direction as HR changes. Here we have again the Saltz (1970) effect of stimulus conditions, which were extremely important in predicting psychophysiological effects.

This argument may also be relevant to the negative findings of Stemmler and Meinhardt (1990), who studied predictions from the general arousal theory. They found that interindividual differences ($N = 42$) in E, N, aggressivity, and somatic complaints (questionnaire data) were related (1) to level and variability of physiological reactivity and (2) to an index of arousability. A large sample of 48 experimental conditions differing widely in "intensity" and emotional "stressfulness," and 34 physiological parameters comprising somatomotor, autonomic, and EEG variables were employed. Physiological data were aggregated into four functionally significant variables. Psychophysiological relationships were analyzed with canonical, multiple, and Pearson correlations. A randomization procedure was used to detect violations of statistical assumptions. Results do not support the predictions: (1) level and variability of reactivity were not related to personality; (2) E was not more closely related to arousability than N; (3) stimulus intensity (physiologically operationalized) failed to be a significant moderator of personality × arousability relationships; and (4) N did not have closer relationships to arousability within emotionally stressful than within nonstressful situations. Results from exploratory analyses suggested that (1) the trait of aggressivity has been hitherto underestimated because it possessed the largest psychophysiological relationships; (2) different sources of arousal (one stimulus bound and one related to the entire setting) could sometimes interact in determining the degree of personality × arousability correlations; and (3) less normative situations (permitting larger behavioral variability) could be at least as effective as highly normative task periods in producing notable correlations. In conclusion, they argue that the arousability concept in personality theory is hampered by the vague notion of what a "situation" is.

Zuckerman (1990) has criticized this study, and the conclusions drawn from it, and Stemmler and Meinhardt (1991) have issued a reply. I feel that experiments involving large numbers of complex scores from psychophysiological experiments and relatively few subjects are inherently unsatisfactory, particularly when the subjects are very homogeneous (female medical students in this case). There are also sequence effects, with the early experiments serving to desensitize the subjects. Single studies controlling conditions of administration in line with theory and previous work are intrinsically more appealing until and unless we have acquired more fundamental knowledge concerning these conditions.

Finally, we may consider a study by T. Ljubin and C. Ljubin (1990). They studied the relation between E and unconditioned reflex reaction to sound— the audiometer reflex (AMR). Audiometer reflex is the motor reflex reaction to a short, intense sound stimulus, appearing in skeletal muscles. It is most obvious in the muscles of the eyelids and face. Since the reflex arc of the AMR reaches the mesencephalic level, AMR is proposed as an indicator of subcortical arousal. An investigation of AMR m.orbicularis oculi was carried out on 12 extraverts, 12 ambiverts, and 12 introverts. Stimulation was accomplished by means of tone intensity of 100 dB, frequency 1000 Hz, duration of 100 ms, binaurally. Reflex response from the AMB m.orbicularis oculi was detected by electromyograph. A statistically significant difference was obtained for the intensity of reflex reaction between introverts and the other two groups (extraverts and ambiverts, which did not differ among themselves), at the .01 level. The difference was in the predicted direction, with introverts showing the highest reflex response amplitude on the average. Results show that AMR amplitude increases with increasing degree of introversion. This was interpreted as a higher subcortical electrophysiological arousal in introverts.

IV. NEUROTICISM AND PSYCHOTICISM

Much less material is available in the field of psychopathological dimensions, even though one might have thought that the preoccupation of psychiatrists and clinical psychologists with N and P would have helped to produce good theories and a great deal of experimental findings. Taking N first, it might seem obvious that high-N subjects react to emotional stimuli earlier, more strongly, and more lastingly than low-N subjects. The literature does not on the whole support such a conclusion although, as already pointed out, one must question the identification of the usual stimuli used as "emotional." The Stroop test, electric shocks or threats thereof, presentation of harrowing slides or films may not produce the kinds of emotional reactions that we normally associate with real-life anxiety, worry, and depression. Equally, certain measurement issues have not been settled (e.g., the application of Wilder's, 1969, Law of Initial Value to electrodermal responses) (Fahrenberg, Walschburger, Foerster, Myrtek, & Muller, 1979).

Oddly enough, early studies by Wenger (1942, 1948, 1957), Theron (1948), van der Merwe (1948), van der Merwe and Theron (1947), Jost (1941), and Freeman (1948) produced quite successful applications of the general theory, which itself dates back to the days of Eppinger and Hess (1917) who advanced the theory of "vagotonia," based on the assumed physiological antagonism of the adrenergic and cholinergic branches of the autonomic nervous system. Predominance of one or the other was thought to explain personality and behavioral differences in emotional responsiveness.

The much more extensive and well-planned studies of Fahrenberg et al (1979) and Myrtek (1980) have essentially failed to show any consistent relationship between autonomic reactivity and measures of anxiety and N. The Fahrenberg study was undertaken on 125 male students who were administered many different psychophysiological tests, as well as personality and other questionnaires. They were also given interviews and asked questions concerning their individual reactions to the various tests. The stress tests used included mental arithmetic under noisy conditions, free speech, interview reactions, and the taking of blood samples. Ten separate psychophysiological measures were recorded: (1) electrodermal activity, (2) an electrocardiogram, (3) blood pressure, (4) pulse frequency and amplitude, (5) skin temperature, (6) a pneumogram, (7) an electromyogram from forehead and arm, (8) eye movement, (9) eyelid movement, and (10) an EEG. The resulting relationships were presented in many detailed tables, and a number of different types of statistical analyses were reported, including correlational analysis, factor analysis, item analysis, multitrait–multimethod analysis, and so on. It is difficult to fault either the experimental or the analytical methods used, although it must be said that there is too little explanation of the rotation methods used for the factor analysis to make judgment of the results possible.

Essentially, the findings of the Fahrenberg et al. and Myrtek studies were completely negative; there were no consistent relationships between different indices of emotionality, or between such indices and personality. Something like eight independent psychophysiological activation factors had to be posited, although it might have been possible through the use of oblique factor rotation to arrive at a more parsimonious result. The Myrtek study gave similar negative results. These two books are invaluable sources of detailed information, and must be regarded as classics.

Some more recent studies have given more positive results. Thus Maushammer, Ehmer, and Eckel (1981) have reported on the relationship between EEG, sensory EPs, and N. Using thirty subjects only, they examined pain thresholds and pain tolerance. N was positively correlated with peak latencies on the sensory EP, correlations depending on the stimulus and intensities used, and ranging from .54 to .73; no correlations were found for E. Harvey and Hirschmann (1980) studied defensive reactions and orienting responses (ORs), following slide presentation of scenes of violent death. Heart responses were recorded, and it was found that initial accelerative responses, indicative of defense, were elicited more frequently from subjects low on E and high on N. In contrast, initial decelerative responses, indicative of orienting, were elicited by subjects scoring high on E and low on N. Results were significant and in good agreement with theory.

Gramer and Huber (in press) have demonstrated the task-specific nature of individual cardiovascular responses to psychological challenge. Using mental arithmetic, anagrams, and preparing a speech on AIDS as the emotion-producing tasks, they recorded HR, systolic (SBP) and diastolic (DBP) blood pressure during pretask and task-exposure periods. Reactivity scores, expressed as difference scores, showed strong task dependence, the tasks showing different patterns of cardiovascular responses. Subjects high on the Emotional Reactivity Scale showed increased SBP and HR reactivity, whereas subjects high on the Social Anxiety Scale demonstrated increased DBP activity. The speech tasks produced the highest correlations.

To make up for these positive sets of results we have the contrary findings of Naveteur and Baque (1987). In their study they examined the relationship between electrodermal activity (EDA) and anxiety in an attention-demanding task. Two hundred sixty one normal volunteer female students completed the Cattell anxiety form, and only subjects with extreme anxiety scores (deciles 0, 1, and 2 on the one hand, $N = 22$; deciles 9 and 10 on the other hand, $N = 24$) were retained. Subjects were presented with a set of 16 stimuli (8 neutral and 8 emotional slides) in a randomized order (different for each subject), of 35-ms and 1-s duration and a randomized interstimuli time averaging 45 s.

Skin conductance levels (SCLs), interstimuli spontaneous fluctuations (SSCRs), skin conductance response (SCR) amplitudes and electrodermal latencies were recorded, as well as skin temperature (ST). Highly anxious subjects showed significantly lower SCLs, lower SCR amplitudes, fewer interstimuli SSCRs, and longer latencies than subjects with low anxiety, whereas ST did not differ between groups. These results go completely counter to theory, with high-anxiety subjects showing less autonomic reactivity. (State anxiety was also measured and gave results more in line with expectation.) Naveteur and Baque list some 50 studies in this field in a summary table that clearly demonstrates the contradictory results achieved, making interpretation very difficult.

To illustrate the difficulties of understanding the directionality of effects when the electrodermal response is involved, consider the following study. B. O. Gilbert and D. G. Gilbert (1991) used movie-induced stress to produce emotional responses, and measured personality on the EPI and MMPI scales. The strong emotion-producing scenes gave larger skin-conductance responses, but all the correlations between personality and SCRs were in the direction of greater emotionality (pathology being associated with *smaller* SCR responses) to the scenes. Marked gender differences made the results even more difficult to explain.

Perhaps the fault lies in a theory that is not specific enough to make precise prediction, unlike the E-arousal theory. Lolas (1987), for instance, has put forward the view that several convergent lines of evidence, both clinical and experimental, suggest that the right hemisphere may be more involved in processing, determining, and expressing emotional stimuli and states. From this, Lolas argues that persons differing in their degree of emotionality may show a differential activation and/or performance of one or the other hemisphere. It was found that N was significantly related to right-hemisphere negative-slope potential amplitudes at central leads; the author interprets this finding as suggesting a relatively higher right-hemisphere activation during a classical fixed four-period RT task. Again, these results are certainly worth following up, but are based on too few subjects to be definitive.

Stenberg (1992) has followed up this possibility. Using students, he instructed his subjects to imagine emotionally laden situations, pleasant or unpleasant, at the same time recording their EEGs. He found that the right-sided fronto-orbital theta rhythm increased in both types of emotion, a finding in line with earlier studies (Bryden & MacRae, 1989; Tucker & Dawson, 1984; Ahern & Schwartz, 1985). The theta factor revealed tonic personality differences, with higher levels in anxious subjects across all conditions. The finding is thus in line with the discovery that regional blood flow is elevated in trait anxious subjects in approximately the same right-sided location (Hagstadius, 1989). Stenberg also found that impulsive individuals showed EEG signs of lower arousal than low impulsives.

Wright, Contrada, and Glass (1985) have suggested the use of more lifelike stimuli to produce emotional behavior, including, for instance, a competition with a $25 prize, between the subject of the experiment and another apparent subject who was really a confederate of the experimenter; he was also an expert on the computer game in question, and practically unbeatable. During the course of the game he would rile the true subject of the experiment, teasing and taunting him. He would also "win" the prize, thus adding injury to insult. The authors report more genuine "emotion" in response to this game than is usually found. But clearly, as already pointed out, the study of neurotic patients gives much closer support to the "limbic system" theory than does the experimental study of "normals," probably because of the unquestioned presence of strong emotions in clinical subjects (Lader & Wing, 1966).

There is little more to add to the story. Although fairly clear deductions can be made from a theory that is widely accepted, these deductions are about as frequently found to be disproved as supported in well-planned studies, and little purpose would be served in listing these contradictory findings. Clearly a more sophisticated theoretical approach is needed, as well as a better way of producing emotional reactions with some degree of certainty. Measurement issues, such as the Law of Initial Value, also require closer attention. Much remains to be done, perhaps following the lines of Thayer's (1989) approach.

If the psychophysiology of N has proved a difficult topic to investigate, biochemical indicators have given more positive results (Bandura, 1991). These effects tend to be related to stress, particularly under conditions that preclude controllability. H. J. Eysenck (1991a) has summarized a large body of evidence showing the relationship between stress-produced emotion (anxiety, depression) and heightened levels of cortisol, endogenous opiates, adrenocorticotropic hormone (ACTH): these in turn are linked with immunodeficiency and cancer. Individual differences contrasted the autonomous behavior of the nonreactors with the neurotic behavior of the other-dependent types; Bandura has used the similar concept of self-efficacy in a way almost synonymous with autonomy. Another point of similarity is that for both authors the crucial argument in favor of a *causal* relation was the successful attempt through some form of psychotherapy, based on learning theory principles, to *alter* the behavior of the neurotic-nonefficient-nonautonomous person in the direction of greater psychological health, with a resulting reversal of the biochemical reaction.

Two studies may be used to illustrate the argument. Bandura et al. (1985) studied severe phobics who were treated by means of a modeling procedure; they were then presented with coping tasks they had previously judged to be in their low, medium, and high self-efficiency range. Continuous blood samples were obtained throughout. Correlations were obtained between self-efficacy beliefs and plasma catecholamine secretion. Epinephrine, norepinephrine (NE), and a DA metabolite, 3, 4-dihydroxyphenylacetic acid (DoPAC) levels were low when phobics coped with tasks in their high-efficiency range. Self-doubts in coping efficacy produced substantial increases in their catecholamines.

Grossarth-Maticek and Eysenck (1989) treated 25 women with terminal cancer of the breast with autonomy-enhancing behavior therapy, while 25 women equally afflicted received no such therapy. (All 50 women received chemotherapy.) The women receiving behavior therapy survived significantly longer, and their lymphocyte percentage increased significantly, while that of the untreated women declined.

There is no room here to review the long list of publications quoted by Bandura (1991) and Eysenck (1991a). Both deal largely with people severely stressed or neurotic, using nonstudent samples; this may be the most impor-tant aspects of successful work in this field. Stress and emotion experienced in these studies are long continued and severe, rather than slight and inter-mittent. Perhaps if psychophysiologists had used similar subject samples they too might have been more successful in their experiments. Unfortunately, few studies combine psychophysiological and biochemical mediators in one de-sign; such a combination would seem likely to throw much light on the relation with personality—even if psychophysiological variables turned out to be nothing but suppressor variables!

One might say that while the theory of E is fairly definite, leading to specific predictions, that of N is less precise, and deductions from it are more difficult to test. As regards P, we have a platform of different theories, all testable, but not joined together under one general heading. Hence our discussion has to be fairly piecemeal, taking one theory at a time. I shall take the most definitive psychophysiological theory first, not only because it is the most directly linked with the topic of this chapter, but also because it has been the most clearly supported. The theory owes its being to Venables (1963) and Claridge (1985; Claridge & Chappa, 1973; Claridge, Robinson, & Birchall, 1985); it asserts that psychosis does not involve a simple shift in, say, emotional arousal, but represents instead a much more complex dissociation of CNS activity. In the schizophrenic, physiological mechanisms that are normally congruent in their activity, thereby maintaining integrated CNS functioning, become uncoupled and dissociated. Some of the empirical work has concentrated on two aspects of CNS functioning considered to be more particularly involved in this un-coupling process, namely emotional arousal, and the mechanisms concerned with the regulation of *sensory input,* including perceptual sensitivity and the broadening and narrowing of attention. Claridge (1981a) called this the phenomenon of reverse covariation.

Specifically, it was proposed that the CNS of the psychotic shows a strong tendency toward altered homeostasis, implicating the mechanism of *arousal* and *selective attention.* Sensory sensitivity might be measured by means of the two-flash threshold, or by examining the degree of "augmenting" and "re-ducing" on the EEG when the intensity of flashes is varied. Generally it is found that sensory responses relate to background arousal in a predictable and intuitively sensible way, but in schizophrenia the opposite is true—positive correlations turn into negative ones. Is what is true of schizophrenics also true of otherwise normal high scorers? This is a crucial test of the continuum hypothesis that I have called the "proportionality criterion" (H. J. Eysenck, 1992); it states that if a test, T_1, derived from hypothesis H, successfully discriminates at a high level between psychotic (particularly schizophrenics) and normals, then T_1, should equally discriminate between high-P and low-P normals, and perhaps also between high-P and low-P psychotics.

The phenomenon of reverse covariation fulfills these conditions admirably. High-P scorers who are not psychotic closely resembled schizophrenics in showing the same profile of psychophysiological response, suggesting a similarity of CNS organization (Claridge, 1985; Claridge & Birchall, 1978; Claridge & Chappa, 1973). The same "psychotic" profile was also shown in "normal" first-degree relatives of schizophrenics, compared with a sample of neurotics' relatives (Claridge et al., 1983, 1985). Robinson and Zahn (1985) have also tested this general hypothesis, with similar results.

A rather different approach is that of Venables (1980) and Flor-Henry and Gruzelier (1983), who postulate a lateralized cerebral dysfunction model. Jutai (1988) supported this hypothesis in his work with schizophrenics, and he also found similar visual search strategies in psychosis-prone (high P) young adults. So did Rawlings and Borge (1987), using the dichotic shadowing technique, and Rawlings and Claridge (1984), using differentially presented visual material. These experiments are too dissimilar to lead to an agreed theoretical statement, but they all agree that high-P subjects (sometimes tested with schizotypy tests rather than P tests) tend to behave like schizophrenics when compared with low-P subjects. Further support for the theory comes from Broeck, Bradshaw, and Szabadi (1991); Hare and Jutai (1988), and Christie and Raine (1988).

Equally relevant and important are a number of studies using measures of hormones, antigens, monoamines, enzymes, and neurotransmitters (Zuckerman, 1991). Thus HLA-B27, a subsystem of the human antigen system is found more frequently in schizophrenics than in normal subjects (Gattaz, Ewald, & Beckman, 1980). Gattaz (1981) has shown that in a comparison of schizophrenic patients with and without HLA-B27, those with the antigen had significantly higher P scores. In another study, Gattaz, Seitz, and Beckman (1985) found that in a nonpsychotic group HLA-B27 positive subjects had significantly higher P scores than negative subjects.

Low platelet monoamine oxidase (MAO) has been found in psychotic patients, and also in their relatives and inpatients who have recovered, suggesting that low MAO activity may be a marker for vulnerability (Buchsbaum, Coursey, & Murphy, 1976; Schalling, Edman, & Asberg, 1987, Schalling et al., 1988). In a recent study of 61 healthy high school volunteers, Klinteberg, Schalling, Edman, Oreland, and Asberg (1987) found correlations of −.30 in females and −.27 in males between MAO and P. It may also be noted that low-MAO activity was found related to E, impulsiveness, and sensation seeking, as well as monotony avoidance, and that Lidberg, Modin, Oreland, Tucker, & Gillner (1985) found it related to psychopathy, again suggesting a relationship between psychopathy and schizophrenia. (See also Checkley, 1980, for a review of MAO in relation to depressive illness.)

These results may be related to serotonin levels, which seem to have similar behavioral correlates as MAO, and hence may be predicted to correlate inversely with P (Zuckerman, 1991). Schalling, Asberg, and Edman (1984) have in fact found that cerebro-spinal fluid (CSF) 5-HIAA levels were inversely related to P scores; similarly, CSF levels of 5-HIAA were found to be positively related to a measure of inhibition of aggression, suggesting that in humans, P is related inversely to the functioning of the serotonergic system, as is much psychopathology.

Theories linking P and the activities of the serotonergic and dopaminergic systems will be found in Pritchard (1991), and the broader theories of schizophrenia of Gray, Feldman, Rawlins, Hemsley, and Smith (1991), Zuckerman (1991), and others reviewed by Eysenck (1992). There is as yet no overall theory linking the partial hypotheses discussed here, but the outlook is promising that in due course P will be firmly anchored in its biological substructure.

V. THE DRUG POSTULATE

One method of studying the psychophysiology of personality is through the use of the drug postulate (Eysenck, 1963). The meaning of this postulate is simply that, given the major dimensions of personality, and given a person's position on these dimensions, this position can be temporarily displaced by the administration of drugs with actions that are relevant to any one of the personality dimensions. An obvious example is alcohol; being a depressant drug it should shift a person's position on the extraversion–introversion axis in the direction of lower arousability (i.e., E). Hallucinogenic drugs should shift a person's position in the direction of higher P, anxiolytic drugs in the direction of greater stability. Figure 12 illustrates the general schema (Eysenck, 1983c).

There are two paradigms for studying the effects of drugs on human personality, as illustrated in Figure 13. Taking as an example the extraversion–introversion dimension, and assuming the essential corrections of the cortical arousal theory, we could take a random sample of the population, tested under placebo condition, and compare their performance on a laboratory task known to be correlated to extraversion–introversion with other members of the random sample given introverting (stimulant) or extraverting (depressant) drugs. The prediction would be that the members of the sample given the *stimulant* drug should behave like introverts in the task, whereas the

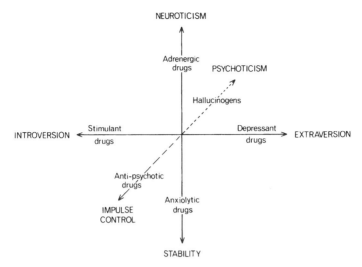

FIGURE 12

The relation between psychotropic drugs and personality. (From H. J. Eysenck, 1983b.)

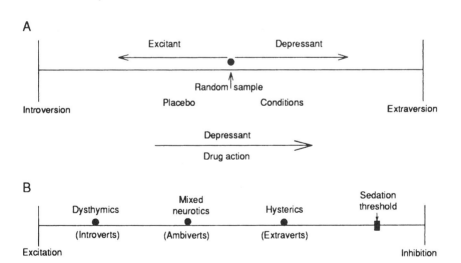

FIGURE 13

Two paradigms for studying the effects of drugs on human personality. (CR, conditioned response.) (From H. J. Eysenck, 1963.)

members of the sample given the depressant drug should behave like extraverts.

As an example, consider eyeblink conditioning. Theory predicts, and experiment confirms, that introverts should condition more quickly than extraverts, with ambiverts in between (H. J. Eysenck & M. W. Eysenck, 1985). Consider now three undifferentiated normal groups given, respectively, a placebo, dexedrene (a stimulant), or amytal (a depressant). The prediction from the arousal theory would be that the dexedrine group would do best, the amytal group worst. Figure 14 shows the results of such a study (Franks & Trouton, 1958). Clearly the prediction is borne out, at a high level of statistical significance.

The alternative method would be to take samples of extraverted, ambiverted and introverted subjects and administer a depressant drug in increasing amounts, until subjects reach a predetermined level of behavior consequent upon the drug administration. A good example here would be the sedation threshold (Shagass, 1954). He defined it in terms of the amount of sodium amytal required to bring about certain behavioral and other changes in the experimental subject, the threshold itself lying somewhat between the state of complete wakefulness and that of complete sedation. One criterion might be the onset of slurred speech, another might be the point on the EEG where inflection occurs in the amplitude curve of induced fast frontal activity. Claridge and Herrington (1963) were the first to provide experimental

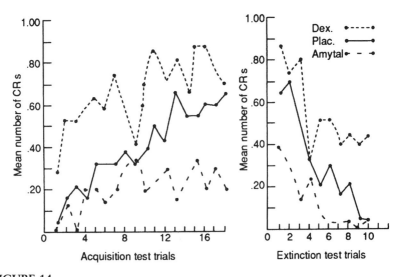

FIGURE 14
Effects of dexedrine and sodium amytal administration on eye-blink conditioning. CR, conditioned response. (From Franks & Trouton, 1958.)

evidence in favor of the deduction that extraverts would require less sodium amytal to reach the threshold than ambiverts, and ambiverts than introverts.

In his book, H. J. Eysenck (1963) referred to 30 successful applications of the drug postulate, and the book itself reported experiments using such techniques as the suppression of the primary visual stimulus (Bidwell's Ghost), visual masking, auditory cross-masking, perceptual inhibition, visual after-images, spiral aftereffect, visual aftereffects, apparent movement, visual con-stancies, pain reactivity, and stress. Later summaries (H. J. Eysenck, 1983a, b) have brought the record up-to-date. Janke (1983) gives an excellent summary of work in this field, demonstrating the complexities of the interactions. Talking about the interactions between N and the drug response, he found that "there was a change according to the drug type, to the dose level, to the situation in which the drug was administered and to the kind of dependent variable" (p. 50). Trouton and Eysenck (1961) have drawn up a table to summarize the complexity of the variables influencing drug effects (Table 1).

One advantage of the general drug postulate is that we can test directly the dimensional hypothesis (e.g., that two so-called depressant drugs have *identical* effects along a single dimension). The consequences are deduced from the assumption that the drugs in question lie at different points on a single continuum, namely the excitation–depression continuum. Consider two de-pressant drugs, doriden and meprobamate. We can test this hypothesis by experimentally creating three groups of subjects (i.e., those tested under doriden, those tested under meprobamate, and those tested under placebo or no-drug conditions). It would be possible to have different people in these three groups, and if the scores on tests used changed considerably due to practice or learning, this would be the only experimental design open to us. However, in view of the great individual differences on test scores even under no-drug conditions, this design would be wasteful and would require large numbers of subjects; consequently a design seems preferable in which the three groups are made up of the same people tested under the three conditions in question. This is the design that was used by various collaborators in H. J. Eysenck's (1960) book *Experiments in Personality,* and data from these ex-periments will be used to illustrate the method of analysis. Five tests in all were used for the analysis, each of which had been employed in experiments involving these three conditions: nonsense syllable learning, RTs, level of skin resistance, flicker fusion, and perimeter threshold differences (H. J. Eysenck and S. B. G. Eysenck, 1960). The choice of the five tests from a much larger battery depended on the total effectiveness of the separation of the groups achieved, and on the necessity of maintaining experimental independence between the tests. A canonical variate analysis of the test scores gave results that are shown in Table 2. It will be seen that the first of the latent roots is significant at the .001 level, whereas the second latent root fails to be significant even at the 5% level. The results therefore bear out the hypothesis

TABLE 1

Some Variables Influencing Drug Effects[a]

I. Nature of the Drug
    A. Preparation (including concentration, vehicle of administration, and whether disguised)
    B. Mode (oral, intravenous injection, etc.) and rate of administration, absorption, and excretion
    C. Dosage (according to body weight or the same for all)
    D. Interval before testing

II. Subject
    A. Personality (including intelligence, extraversion–introversion, neuroticism, suggestibility, etc.)
    B. Familiarity with the situation and the amount of stress occasioned by it
    C. Practice, fatigue, motivation
    D. Tendency to react to placebos
    E. Psychiatric state and its duration
    F. Age, sex, physics, height, and weight
    G. Present state
    H. General state of health, nutritional status, sleep
    I. Conditions of work (e.g., temperature, humidity, oxygen, lack of)
    J. Diseases, disabilities (e.g., fever, thyrotoxicosis, liver or kidney damage), or effects of operations, etc. (e.g., leucotomy, concussion)
    K. Time of day
    L. Interval since last meal (if drug given orally)
    M. Recent medication with other drugs (e.g., sedation) or ingestion of drinks containing stimulants or depressants
    N. Previous experience of drugs
    O. Cumulative effect of some drugs (e.g., bromides)
    P. Habituation, tolerance (including cross-tolerance)
    Q. Addiction
    R. Idiosyncracy of hypersensitivity

III. Social Environment
    A. Interaction with other subjects
    B. Activities required or permitted after administration of the drug
    C. Suggestion
    D. Reinforcement of responses by the experimenter

---

[a]From Trouton and Eysenck, 1961.

and show that both meprobamate and doriden lie on a continuum with respect to their psychological effects, and do not differ from each other in mode of action. (The term "mode of action" does not of course refer to the biochemical and physiological type of action, but merely to the behavioral effects of these drugs.) It is clearly possible that had other tests been used, results might have been different, but the experiment will suffice to illustrate the method.

TABLE 2

Canonical Variate Analysis of Five Tests, Showing Effects of Two Drugs

Latent vectors	X_1	X_2
1. Nonsense syllable learning	−.040971	−.014399
2. Reaction time	−.013625	−.161517
3. Level of skin resistance	−.080835	−.013827
4. Flicker fusion	1.000000	1.000000
5. Perimeter threshold difference	−.475330	.147773

Latent roots
   $\lambda_1 = .453220 = 90.03\%$ $\lambda_2 = .050205 = 9.97\%$
   Diagonal entries of matrix $G^{-1}B = 0.503425$

Significance of roots
   $R_1^2 = \lambda_1 : X^2 = 40.446 : p < 0.001$
   $R_2^2 = \lambda_2 : X^2 = 3.450 :$ n.s.

[a]From Eysenck and Eysenck, 1960.

Scores were computed for each subject on both canonical variates, and the means of these scores were used to calculate what might be called the percentile of correct classification, (i.e., given that we only know the canonical variate scores of a person, could we correctly identify him as a person who had been tested under no-drug condition, under meprobamate, or under doriden?) Results of applying this method showed that over 70% of accurate classifications could be made under these conditions, which is a remarkable achievement considering (1) the very slight amount of drug administered, and (2) the fact that both drugs had the same effect, and were only differentiated by *strength* of the effect. (Nearly all the discrimination is of course contributed by the first value; the second one might have been omitted without much change in the number of misclassifications.) This approach could of course have been greatly enlarged by having stimulant as well as depressant drugs, by having further tests, and finally by extending it to other dimensions as well, either in separate or in identical designs. It is suggested that the drug postulate could thus be quantified very much more extensively and accurately than is possible with single experiments employing only one laboratory measure. Such a task clearly lies in the future.

A large number of more recent studies using the drug postulate have been discussed elsewhere (H. J. Eysenck and S. B. G. Eysenck, 1985); the general outcome has usually been to support the theoretical model, but often with modification. Thus the sedation threshold type of experiment usually demonstrates complex interaction between E and N (Claridge, Donald, and Birchall, 1981). In view of the complexity of drug experiments suggested in Table 1, it will not come as a surprise that many experimental findings, although usually in the right direction, throw up certain associations that require further investigation.

Perhaps the most direct evidence linked with theory has been in the field of vigilance, where the correlation with I is perhaps clearest. Here O'Hanlon (1965) found that the concentration of adrenalin (a substance associated with arousal) decreased during a very long vigilance task among subjects who showed a vigilance performance decrement, but not among those whose performance remained stable over time. Furthermore, the correlation between target detection and adrenalin concentration was .84 among those manifesting a decrement in performance. Perhaps this study will be considered to have inaugurated a third paradigm of personality–drug interaction.

In recent years relatively few applications of the drug postulate have been published, usually without awareness of its existence. Nevertheless the results have been of considerable interest. Only a few will be discussed here. In one of these, Haier, Reynolds, Prager, and Cox (1991) studied the influence of flurbiprofen and caffeine, in interaction with E, as analgesics.

Normal volunteers received the analgesic drug flurbiprofen (50 mg) with or without caffeine (100 mg) in a placebo-controlled study. Pain sensitivity was determined as response to electric shock stimuli using a signal detection paradigm at 30-min intervals for 2 hr. A three-way ANOVA showed a significant analgesic effect for the drug but no greater or faster effect was apparent in the drug-plus-caffeine combination. When the subjects were divided into introverts and extraverts, caffeine increased drug analgesia for both groups but at different stimulus intensities; lower for introverts and higher for extraverts. The data suggest that introverts who take analgesics in combination with caffeine may even experience potentiation of pain stimulation. The findings support the importance of arousal in pain perception, and also demonstrate predictable effects of caffeine–introversion interaction.

Another study is relevant to the theories of psychoticism discussed in the previous section. Netter and Rammsayer (1991) studied the effect of dopaminergic drugs on aggression-related personality traits. In a double-blind crossover experiment the personality dependent effects of a DA antagonist (haloperidol) and a DA precursor (L-dopa) on vigilance, psychomotor performance, and subjective activation was investigated in 24 healthy male volunteers divided according to questionnaire scores into high and low aggressive or high and low scorers of the sensation-seeking subscales. Correlations and analyses of variance were performed on drug-induced change scores. It was concluded that the personality score of aggressiveness as well as boredom susceptibility as one of the aspects of aggression seemed to be related to a reduction of performance and increase in tension by L-dopa and either no effect or an increase in stability of performance by haloperidol. In high-thrill and adventure-seeking (TAS) subjects a similar resistance to effects of haloperidol became evident in body temperature, which remained unchanged, whereas it was decreased in low-TAS subjects. It was concluded that aggressiveness and its subaspects are related to lower sensitivity of DA receptors pointing to a biological DA-related dimension ranging from aggression in the normal population to schizophrenia.

An implication of the drug postulate is that what is important is the relevant drug status of the individual, which can be altered by the administration of drugs, but also by external events; thus cortisol level is affected by stress that is regarded as uncontrollable (H. J. Eysenck, 1991a). A study by Dabbs and Hopper (1990) considered this paradigm. The relationship of the adrenal hormone cortisol to arousal and personality was examined among 102 male college students and 4462 male military veterans. Students high in cortisol had high HRs, and questionnaire scores showed them to be anxious and depressed, solitary, and appreciative of fantasy, aesthetics, ideas, and values. Veterans high in cortisol had high HR and high SPB and DPB, and Minnesota Multiphasic Personality Inventory (MMPI) scores showed them to be anxious, depressed, introverted, masculine, and not manic. Results were consistent with the notion that high levels of cortisol are associated with high arousal, which is debilitating and leads subjects to avoid the further arousal that would come from encounters with novel persons and events. However, the sizes of the effects were small, especially among MMPI scores for the veterans.

In a similar study Dabbs, Hopper, and Jurkovic (1990) looked at testosterone (T), which had been linked with P (H. J. Eysenck and S. B. G. Eysenck, 1976). Relationships of serum and saliva to personality were examined among 401 college students in four laboratory studies and 5236 military veterans in one archival study. Among the students, there were few relationships between T and traditional personality measures. Among the veterans, MMPI scores and *Diagnostic and Statistical Manual of Mental Disorders* (*DSM-III*) diagnoses showed T related to drug and alcohol abuse, antisocial and generally intemperate behavior, and affective disorders. Consistent with social control theory, correlations were higher among veterans who were lower in socioeconomic status. It appears likely that T has innate effects that are socially undesirable but can be attenuated by bonds between the individual and society. Effect sizes were small, suggesting that T will have noticeable effects only in large populations or individuals who differ markedly from the population mean. These results are in good agreement with theory.

It is curious that psychophysiological studies of personality have flourished, in spite of the largely negative findings in relation to N, whereas drug studies, which have a much higher level of success, have not. Perhaps the need for medical supervision when drugs are used has contraindicated their use for psychologists interested in the study of personality. Whatever the cause, it seems a pity that so little use has been made of an experimental approach that goes beyond simple correlation, and promises a proper causal type of analysis. Theoretically, the drug postulate seems to hold out the hope of unifying many divergent strands of research in the field of biological personality study.

VI. CONCLUSIONS

As always in summarizing scientific research in psychology, we may look at the glass as half-full or half-empty. My own conclusion is probably more optimistic than that of many other reviewers of the evidence. Given the enormous complexity of the issues involved, the practical difficulties outlined, and the instrumental problems of working at the boundary of the technically possible, I think we would have been optimistic in the extreme had we expected results to have been more positive and in greater agreement with theory. Inevitably early theories in any field are fuzzy, disjointed, and imprecise; it is only by being brought in contact with attempts to test deductions from such theories that they become more precise and realistic. By virtue of the research reported here we know a great deal more than we did 50 years ago; surely, that is a gain?

Theoretical advances usually proceed along several stages. The first stage confronts immature theories with attempts at deduction testing that cannot specify the optimum conditions for testing, with the results that any successful outcome is accompanied by several unsuccessful ones—like Newton's failures to predict or account for the movements of the moon. Gradually we begin to obtain a better grasp of the conditions that are required to test our theories more appropriately, and enter the second stage. We begin to avoid testing in situations that are inappropriate, either by being too boring or by being too stressful; we learn about stimulus properties that give the best results; we discover the best response measures to use, as well as optimal electrode placements. Gradually we move into the third stage, where personality theories and testing conditions are seen as part of a single scheme or system leading to a higher level of integration. This in turn leads to the fourth stage where we finally arrive at a proper paradigm of personality.

Clearly we have not reached the higher levels or stages yet. I would guess that we are halfway in the second stage, within a reasonable distance of it and with a high degree of likelihood of reaching it in the not too distant future. What has held progress back has been the undisciplined mode of attack on the many problems involved. Clearly the Yerkes-Dodson law is vitally involved in our studies, as is Pavlov's law of transmarginal inhibition. Yet hardly anyone attacks the questions involved directly; we do not even know if the two laws are in fact identical, or refer to different types of stimulation and task properties. Instead of attacking fundamental problems, thousands of personality psychologists make up useless new psychometric tests that serve no theoretical purpose and will not advance our scientific knowledge one iota. Science is built brick by brick, one person building on the contributions of others; psychology, and the study of personality in particular, is more like people throwing bricks at each other—no wonderful new buildings will eventuate in this fashion. We know enough to suggest that we can discover the biological roots of our personality; the glass is half-full, and we have the chance to fill it completely if we imitate the hard sciences and work according to a plan, instead of pulling in all directions.

REFERENCES

Ahern, G. L., & Schwartz, G. E. (1985). Differential lateralization for positive and negative emotion in the human brain: EEG spectral analysis. *Neuropsychologia, 23,* 745–755.

Anderson, K. J. (1990). Arousal of the inverted-U hypothesis: Nine critiques of Neiss's "Reconceptualizing Arousal." *Psychological Bulletin, 107,* 96–100.

Andress, D. L., & Church, M. W. (1981). Differences in brainstem auditory evoked responses between introverts and extraverts as a function of stimulus intensity. *Psychophysiology, 18,* 156.

Bandura, A. (1991). Self-efficacy mechanisms in physiological activation and health-promoting behavior. In John Modden (Ed.), *Neurobiology of Learning, Emotion and Affect* (pp. 229–269). New York: Raven Press.

Bandura, A., Taylor, C. B., Williams, S. L., Mefford, I. N., & Barshas, J. P. (1985). Catecholamine secretion as a function of perceived coping self-efficacy. *Journal of Consulting and Clinical Psychology, 53,* 406–414.

Bartussek, D., Diedrich, O., Naumann, E., & Collet, W. (1993). Introversion–extraversion and event-related potential (ERP): A test of J. A. Gray's theory. *Personality and Individual Differences, 14,* 565–574.

Bartussek, D., Naumann, E., Collet, W., & Moeller, H. (1990). *Ereigniskorreliertes Hirnrindenpotential (EKP) in Abhangig-keit von der Rueckmeldebedeutung der Reize, Extraversion und Neurotizismus.* Trier: Trierer Psychologische Berichte, *17,* 8, 1–56.

Bartussek, D., Naumann, E., Moeller, H., Vogelbacher, D., & Diedrich, O. (1990). Emotionale Wortbedeutung, Extraversion und der spaete positive Komplex des Ereigniskorrelierten Potentials (EKP). Lecture given by O. Diedrich at the *37th Congress of the German Psychological Society* in Kiel, Germany.

Berlyne, D. E. (1960). *Conflict, arousal and curiosity.* New York: McGraw Hill.

Birbaumer, N., & Schmidt, R. E. (1991). *Biologische Psychologie, 2nd Ed.* New York: Springer-Verlag.

Blake, M. J. F. (1967). Relationship between circulation rhythm of body temperature and introversion–extraversion. *Nature, 215,* 896–897.

Bouchard, T. J. (1991). A twice-bold tale: Twins reared apart. In W. M. Grove & D. Cicchetti (Eds.), *Personality and Psychopathology* (pp. 188–215). Minneapolis: University of Minnesota Press.

Brebner, J., & Cooper, C. (1978). Stimulus or response-produced excitation: A comparison of the behavior of introverts and extraverts. *Journal of Research in Personality, 12,* 306–311.

Britt, T. W., & Blumenthal, T. D. (1991). Motorneuronal insensitivity in extraverts as revealed by the startle response paradigm. *Personality and Individual Differences, 12,* 387–393.

Broadhurst, P. L. (1959). The interaction of task difficulty and motivation: The Yerkes-Dodson law revived. *Acta Psychologia, 10,* 321–338.

Broeck, M. O. van, Bradshaw, L. M., & Szabadi, E. (1991). The relationship between "impulsiveness" and hemispheric functional asymmetry, investigated with a divided visual field word recognition task. *Personality and Individual Differences, 13,* 355–360.

Brunean, W., Roux, S., Perse, S., & Lelord, G. (1984). Frontal evoked responses, stimulus intensity control, and the extraversion dimension. *Annals of the New York Academy of Sciences, 425,* 546–550.

Bryden, M. P., & MacRae, L. (1989). Dichotic laterality effects obtained with emotional words. *Neuropsychiatry, Neuropsychology and Behavioral Neuroscience, 1,* 71–76.

Buchsbaum, M. S., Coursey, R. O., & Murphy, D. L. (1976). The biochemical high-risk paradigm: Behavioral and familial correlates of low platelet monoamine oxidase activity. *Science, 194,* 335–341.

Buchsbaum, M. S., & Haier, R. J. (1983). Individual differences in augmenting-reducing evoked potentials. In A. Gale & J. Edwards (Eds.), *Physiological correlates of human behavior, Vol. 3.* London: Academic Press.

Caccioppo, J. T., & Tassinary, L. G. (Eds.) (1990). *Principles of Psychophysiology: Physical, Social and Inferential Elements.* New York: Cambridge University Press.

Campbell, K. B., Barikean-Braun, J., & Braun, C. (1981). Neuroanatomical and physiological foundations of extraversion. *Psychophysiology, 18,* 263–267.

Checkley, S. A. (1980). Neuroendocrine tests of monoamine function in man: A review of basic theory and its application to the study of depressive illness. *Psychological Medicine, 10,* 35–53.

Claridge, G. S. (1981a). Psychoticism. In R. Lynn (Ed.), *Dimensions of personality* (pp. 79–109). Oxford: Pergamon Press.

Claridge, G. S. (1981b). The Eysenck psychoticism scale. In J. P. Butcher & C. D. Spielberger (Eds.), *Advances in Personality Assessment, Vol. 2* (pp. 71–114). Hillsdale, NJ: Erlbaum.

Claridge, G. S. (1985). *Origins of mental illness.* Oxford: Blackwell.

Claridge, G. S., & Birchall, P. M. J. (1978). Bishop, Eysenck, Block and psychoticism. *Journal of Abnormal Psychology, 87,* 604–668.

Claridge, G. S., & Chappa, H. J. (1973). Psychoticism: A study of its biological nature in normal subjects. *British Journal of Social and Clinical Psychology, 12,* 175–187.

Claridge, G. S., Donald, J. R., & Birchall, P. M. (1981). Drug tolerance and personality: Some implications for Eysenck's theory. *Personality and Individual Differences, 2,* 153–166.

Claridge, G. S., & Herrington, R. N. (1963). Excitation-inhibition and the theory of neurosis: A study of the sedatic threshold. In H. J. Eysenck (Ed.), *Experiments with Drugs* (131–168).

Claridge, G., Robinson, D. L., & Birchall, P. (1983). Characteristics of schizophrenics' and neurotics' relatives. *Personality and Individual Differences, 4,* 651–664.

Claridge, G., Robinson, D. L., & Birchall, P. (1985). Psychophysiological evidence of "psychoticism" in schizophrenics' relatives. *Personality and Individual Differences, 6,* 1–10.

Christie, M., & Raine, A. (1988). Lateralized hemisphere activity in relation to personality and degree course. *Personality and Individual Differences, 9,* 957–964.

Cohen, J. (1988). *Statistical power analysis for the behavioral sciences.* (2nd Ed.). Hillsdale, NJ: Erlbaum.

Cohen, J. (1990). Things I have learned (so far). *American Psychologist, 75,* 1304–1312.

Cohen, M. R., & Nagel, N. (1936). *An introduction to logic and scientific method.* New York: Harcourt & Brace.

Coles, M. G. H., Donchin, E., & Porges, S. W. (1986). *Psychophysiology: Systems, processes, and applications.* New York: Guilford Press.

Cronbach, L. J. (1957). The two disciplines of scientific psychology. *American Psychologist, 12,* 671–684.

Crow, T. J. (1986). The continuum of psychosis and its implication for the structure of the gene. *British Journal of Psychiatry, 149,* 419–429.

Dabbs, J. M., & Hopper, C. H. (1990). Cortisol, arousal, and personality in two groups of normal men. *Personality and Individual Differences, 11,* 931–935.

Dabbs, J. M., Hopper, C. H., & Junkovic, G. J. (1990). Testosterone and personality among college students and military veterans. *Personality and Individual Differences, 11,* 1203–1269.

Daruna, J. H., Karrer, R., & Rosen, A. J. (1985). Introversion, attention and the late positive component of event-related potentials. *Biological Psychology, 20,* 249–259.

Deary, I. J., Ramsay, H., Wilson, J. A., & Riad, M. (1988). Stimulated salivation: Correlations with personality and time of day effects. *Personality and Individual Differences, 9,* 903–909.

Dincheva, E., & Piperova-Dalbokova, D. C. (1982). Differences in contingent negative variation (CNV) related to extraversion–introversion. *Personality and Individual Differences, 3*, 447–451.

Dincheva, E., Piperova-Dalbokova, D., & Kolev, P. (1984). Contingent negative variation (CNV) and the distraction effect in extraverts and introverts. *Personality and Individual Differences, 5*, 757–761.

Di Traglia, G., & Polich, J. (1991). P300 and introverted/extraverted personality types. *Psychophysiology, 28*, 177–184.

Donchin, E., & Coles, M. G. (1988). Is the P300 component a manifestation of content updating? *Brain and Behavioral Sciences, 11*, 357–374.

Eaves, L. J., Eysenck, H. J., & Martin, N. G. (1989). *Genes, culture and personality: An empirical approach.* San Diego: Academic Press.

Eppinger, H., & Hess, W. R. (1917). Vagotonia. *Nervous and Mental Disease Monographs, 20.*

Eysenck, H. J. (1950). Criterion analysis—an application of the hypothetico-deductive method to factor analysis. *Psychological Review, 57*, 38–53.

Eysenck, H. J. (1952a). *The scientific study of personality.* London: Routledge & Kegan Paul.

Eysenck, H. J. (1952b). Schizothymia-Cyclothymia as a dimension of personality: II. Experimental. *Journal of Personality, 20*, 345–384.

Eysenck, H. J. (1957). *The dynamics of anxiety and hysteria.* London: Routledge & Kegan Paul.

Eysenck, H. J. (1960). The place of theory in psychology. In H. J. Eysenck (Ed.), *Experiments in Personality, Vol. 2* (303–315). London: Routledge & Kegan Paul.

Eysenck, H. J. (Ed.) (1963). *Experiments with drugs.* Oxford: Pergamon.

Eysenck, H. J. (1967). *The biological basis of personality.* Springfield, C. C. Thomas.

Eysenck, H. J. (1976). *Sex and personality.* London: Open Books.

Eysenck, H. J. (Ed.) (1981). *A Model for Personality.* New York: Springer.

Eysenck, H. J. (1983a). Is there a paradigm in personality research? *Journal of Research Personality, 17*, 369–397.

Eysenck, H. J. (1983b). Drugs as research tools in psychology: Experiments with drugs in personality research. *Neuropsychobiology, 10*, 29–43.

Eysenck, H. J. (1983c). Psychopharmacology and personality. In W. Janke (Ed.), *Response variability to psychotropic drugs* (127–154). Oxford: Pergamon Press.

Eysenck, H. J. (1984). The place of individual differences in a scientific psychology. In J. R. Royce and L. P. Mos (Eds.), *Annals of Theoretical Psychology, Vol. 1* (233–314). New York: Plenum Press.

Eysenck, H. J. (1985). The place of theory in a world of facts. In K. B. Madsen & L. P. Mos (Eds.), *Annals of Theoretical Psychology, Vol. 3* (17–114).

Eysenck, H. J. (1990). Biological dimensions of personality. In L. A. Pervin (Ed.), *Handbook of personality* (244–276). New York: Guilford Press.

Eysenck, H. J. (1991a). *Smoking, personality and stress: Psychosocial factors in the prevention of cancer and coronary heart disease.* New York: Springer Verlag.

Eysenck, H. J. (1991b). Dimensions of personality: 16, 5 or 3?—Criteria for a taxonomic paradigm. *Personality and Individual Differences, 12*, 773–790.

Eysenck, H. J. (1992). The definition and measurement of psychoticism. *Personality and Individual Differences, 13*, 757–785.

Eysenck, H. J. (1993). Creativity and personality: Suggestions for a theory. *Psychological Inquiry, 4*, 147–246.

Eysenck, H. J., & Eysenck, M. W. (1985). *Personality and individual differences.* New York: Plenum Press.

Eysenck, H. J., & Eysenck, S. B. G. (1960). The classification of drugs according to their behavioral effects: A new method. In H. J. Eysenck (Ed.), *Experiments in Personality, Vol. 1* (225–233). London: Routledge & Kegan Paul.

Eysenck, H. J., & Eysenck, S. B. G. (1976). *Psychoticism as a Dimension of Personality.* London: Hodder & Stoughton.

Eysenck, S. B. G., & Eysenck, H. J. (1967a). Salivary response to lemon juice as a measure of introversion. *Perceptual & Motor Skills, 24,* 1047–1053.

Eysenck, S. B. G., & Eysenck, H. J. (1967b). Physiological reactivity to sensory stimulation as a measure of personality. *Psychological Reports, 20,* 45–46.

Fahrenberg, J., Walschburger, P., Foerster, M., Myrtek, M., & Muller, W. (1979). *Psychophysiologische Aktivierungsforschung.* Munich: Minerva.

Flor-Henry, P., & Gruzelier, J. (1983). *Laterality of Psychopathology.* New York: Elsevier.

Franks, C. M., & Trouton, D. S. (1958). Effects of lamobarbistal sodium and dexamphetamine sulphate on the conditioning of the eyeblink response. *Journal of Comparative and Physiological Psychology, 51,* 220–222.

Freeman, G. L. (1948). *The Energetics of Human Behavior.* Ithaca, NY: Cornell University Press.

Gale, A. (1983). Electroencephalographic studies of extraversion–introversion: A case study in the psychophysiology of individual differences. *Personality and Individual Differences, 4,* 371–380.

Gale, A. (1984). O'Gorman versus Gale: a reply. *Biological Psychology, 19,* 129–136.

Gale, A., & Eysenck, M. W. (Eds.) (1992). *Handbook of Individual Differences: Biological Perspective.* New York: Wiley.

Gattaz, W. F. (1981). HLA-B27 as a possible genetic marker of psychoticism. *Personality and Individual Differences, 2,* 57–60.

Gattaz, W. F., Ewald, R. V., & Beckman, H. (1980). The HLA system and schizophrenia. *Archives fur Psychiatrie und Nervenkrankheiten, 228,* 205–211.

Gattaz, W. F., Seitz, M., & Beckman, H. (1985). A possible association between HLA 13-27 and vulnerability to schizophrenia. *Personality and Individual Differences, 6,* 283–285.

Geen, R. G. (1984). Preferred stimulation levels in introverts and extroverts: Effects of arousal and performance. *Journal of Personality and Social Psychology, 46,* 1303–1312.

Gilbert, B. O. (1991). Physiological and nonverbal correlates of extraversion, neuroticism, and psychoticism during active and passive coping. *Personality and Individual Differences, 12,* 1325–1331.

Gilbert, B. O., & Gilbert, D. G. (1991). Electrodermal responses to movie-induced stress as a function of EPI and MMPI scale scores. *Journal of Social Behavior and Personality, 6,* 903–914.

Golding, J. F., & Richards, M. (1985). EEG spectral analysis, visual evoked potential and photic-driving correlates of personality and memory. *Personality and Individual Differences, 6,* 67–76.

Goode, D. J., & Manning, A. A. (1985). Variation in the Hoffman reflex recovery curve related to clinical manifestations of schizophrenic disorder. *Psychiatric Research, 15,* 63–70.

Gramer, M., & Huber, H. (in press). Individual variability in task-specific cardiovascular response patterns during psychological challenge. *Personality and Individual Differences.*

Gray, J. (1981). A critique of Eysenck's theory of personality. In H. J. Eysenck (Ed.), *A model for personality.* New York: Springer.

Gray, J. A., Feldman, J., Rawlins, J. P., Hemsley, D. D., & Smith, A. D. (1991). The neuropsychology of schizophrenia. *Behavioral and Brain Science, 14,* 1–84.

Greenaway, F. (1966). *John Dalton and the atom.* London: Heinemann.

Grossarth-Maticek, R., & Eysenck, H. J. (1989). Length of survival and lymphocyte percentage in women with mammary cancer as a function of psychotherapy. *Psychological Reports, 65,* 315–321.

Hagstadius, S. (1989). *Anxiety proneness related to the regional cerebral blood flow during resting and mental activation.* Unpublished doctoral dissertation, quoted by Stenberg, in press.

Haier, F., & Hirschmann, R. (1980). The influence of extraversion and neuroticism on heart rate responses to aversive visual stimuli. *Personality and Individual Differences, 1,* 97–100.

Haier, R. J., Reynolds, C., Prager, E., Cox, S., & Buchsbaum, M. S. (1991). Flurbiprofen, caffeine and analgesia: Interaction with introversion/extraversion. *Personality and Individual Differences, 12,* 1349–1354.

Haier, R. J., Robinson, D. L., Bradley, W., & Williams, D. (1984). Evoked potential augmenting-reducing of personality differences. *Personality and Individual Differences, 5,* 293–301.

Hare, R. D., & Jutai, J. W. (1988). Psychopathy and cerebral asymmetry in semantic processing. *Personality and Individual Differences, 9,* 329–337.

Houston, B. K. (1986). Psychological variables and cardiovascular and neuroendocrine reactivity. In K. A. Matthews, S. M. Weiss, T. Detre, T. M. Dembrowski, B. Falkner, S. M. Manuck, & R. B. Williams, (Eds.), *Handbook of Stress, Reactivity, and Cardiovascular Disease* (207–230). New York: Wiley.

Ingwar, D. H., & Risberg, J. (1965). Influence of mental activity upon regional cerebral blood flow in men. *Acta Neurologica Scandinavica, 41,* (Suppl. 14), 43–96.

Ingwar, D. H., & Risberg, J. (1967). Increase of regional cerebral blood flow during mental effort in normals and in patients with focal brain disorders. *Experimental Brain Research, 3,* 155–211.

Janke, W. (Ed.) (1983). *Response variability to psychotropic drugs.* Oxford: Pergamon Press.

John, O. (1990). The "Big Five" factor taxonomy. In L. A. Pervin (Ed.), *Handbook of Personality* (66–100). New York: Guilford Press.

Jost, H. (1941). Some physiological changes during frustration. *Child Development, 12,* 9–15.

Jutai, J. W. (1988). Spatial attention in hypothetically psychosis-prone college students. *Psychiatry Research, 27,* 207–215.

Kaada, B. R. (1960). Neurophysiology. In J. Field and H. W. Magom (Eds.), *Handbook of Physiology* (135–1372). Washington, D.C.: American Physiological Society.

Kimble, G. A. (1961). *Hilgard & Marquis' conditioning and learning.* New York: Appleton-Century-Crofts.

Klinteberg, B., Schalling, D., Edman, G., Oreland, L., & Asberg, M. (1987). Personality correlates of platelet monoamine oxidase (MAO) activity in female and male subjects. *Neuropsychobiology, 18,* 89–96.

Koelega, H. S., & Verbaten, M. N. (1991). Event-related brain potentials and vigilance performance: Dissociations abound, a review. *Perceptual and Motor Skills, 72,* 971–982.

Kuhn, T. S. (1970). *The structure of scientific revolutions.* Chicago: University of Chicago Press.

Lacey, J. I. (1950). Individual differences in somatic response patterns. *Journal of Comparative and Physiological Psychology, 43,* 338–350.

Lacey, J. I. (1967). Somatic response patterning and stress: Some revisions of activation theory. In M. H. Appleby & R. Turnbull (Eds.), *Psychological Stress.* New York: Appleton-Century-Crofts.

Lacey, J. I., & Lacey, B. C. (1958). Verification and extension of the principle of autonomic response-stereotype. *American Journal of Psychology, 71,* 50–73.

Lader, M. (1973). The psychophysiology of hysterics. *Journal of Psychosomatic Research, 17,* 265–269.

Lader, M., & Wing, L. (1966). *Physiological measures, sedative drugs and morbid anxiety.* Oxford: Maudsley Monograph No. 14.

Lakatos, I., & Musgrave, A. (Eds.) (1970). *Criticism and the growth of knowledge.* Cambridge: Cambridge University Press.

Lidberg, L., Modin, I., Oreland, L., Tucker, J. R., & Gillner, A. (1985). Platelet monoamine oxidase activities and psychopathy. *Psychiatry Research, 16,* 339–343.

Ljubin, T., & Ljubin, C. (1990). Extraversion and audiometer reflex. *Personality and Individual Differences, 11,* 977–984.

Lolas, F. (1987). Hemispheric asymmetry of slow brain potentials in relation to neuroticism. *Personality and Individual Differences, 8,* 969–971.

Lolas, F., & Andrassa, I. (1977). Neuroticism and extraversion and slow brain potentials. *Neuropsychobiology, 3,* 12–22.

Lukas, J. N. (1987). Visual evoked potential augmenting-reducing and personality: The vertex augmenter is a sensation-seeker. *Personality and Individual Differences, 8,* 385–395.

Magliero, A., Bashore, T., Coles, M. G., & Donchin, E. (1989). On the dependence of the P300 latency on stimulus evaluation processes. *Psychophysiology, 21,* 171–186.

Mangan, G. (1982). *The biology of human conduct.* Oxford: Pergamon Press.

Mangan, G. L., Pellett, D., Colrain, I. M., & Bates, T. C. (1993). Photic driving and personality. *Personality and Individual Differences.*

Matthews, G., & Amelang, M. (1992). Extraversion, arousal theory and performance: A study of individual differences in the EEG. *Personality and Individual Differences, 14,* 347–363.

Matthews, R. J., Weinman, M. L., & Barr, D. L. (1984). Personality and regional cerebral blood flow. *Bristol Journal of Psychiatry, 144,* 524–532.

Maushammer, C., Ehmer, G., & Eckel, K. (1981). Pain, personality and individual differences in sensory evoked potentials. *Personality and Individual Differences, 2,* 335–336.

Myrtek, M. (1980). *Psychophysiologische Konstitutionsforschung.* Gottingen, West Germany: Hogrefe.

Naveteur, L. J., & Baque, E. F. (1987). Individual differences in the electrodermal activity as a function of subjects' anxiety. *Personality and Individual Differences, 8,* 615–626.

Neiss, R. (1988). Reconceptualizing arousal: Psychobiological study in motor performance. *Psychological Bulletin, 103,* 345–366.i

Netter, P., & Rammsayer, T. (1991). Reactivity to dopaminergic drugs and aggression-related personality traits. *Personality and Individual Differences, 12,* 1009–1017.

Neubauer, A. C. (1992). Psychosomatic comparison of two circadian rhythm questionnaires and their relationships with personality. *Personality and Individual Differences, 13,* 125–131.

O'Connor, K. (1983). Individual differences in components of slow cortical potentials: Implications for models of information processing. *Personality and Individual Differences, 4,* 403–410.

O'Gorman, J. G. (1984). Extraversion and the EEG: I. An evaluation of Gale's hypothesis. *Biological Psychology, 19,* 95–112.

O'Gorman, J. G., & Lloyd, J. E. M. (1987). Extraversion, impulsiveness, and EEG alpha activity. *Personality and Individual Differences, 8,* 169–174.

O'Gorman, J. G., & Mallise, L. R. (1984). Extraversion of the EEG: II. A test of Gale's hypothesis. *Biological Psychology, 15,* 113–127.

O'Hanlon, J. (1965). Adrenaline and noradrenaline: Relation to performance in a visual vigilance task. *Science, 150,* 507–509.

Orlebeke, J. F., Kok, A., & Zeillensaker, G. W. (1989). Distribution and the processing of auditory stimulus intensity: An ERP study. *Personality and Individual Differences, 10,* 445–457.

Ortiz, T., & Maojo, V. (1993). Comparison of the P300 wave in introverts and extraverts. *Personality and Individual Differences.*

Pascalis, V. de, & Montirosso, R. (1988). Extraversion, neuroticism and individual differences in event-related potentials. *Personality and Individual Differences, 9,* 353–360.

Pearson, G. L., Freeman, F. G. (1991). Effects of extraversion and mental arithmetic on heart-rate reactivity. *Perceptual and Motor Skills, 72,* 1235–1244.

Piperova-Dalbokova, D., Dincheva, E., & Urgelles, L. (1984). Stability of contingent negative variation (CNV) and extraversion–introversion. *Personality and Individual Differences, 5,* 763–766.

Pivik, R. T., Stelmack, R. M., & Blysma, F. W. (1988). Personality and individual differences in spiral motorneuronal excitability. *Psychophysiology, 25,* 16–24.

Plorij-van Gorsel, E. (1981). EEG and cardiac correlates of neuroticism: A psychophysiological comparison of neurotics and normal controls in relation to personality. *Biological Psychology, 13,* 141–156.

Plorij-van Gorsel, E., & Janssen, R. (1978). Contingent negative variation (CNV) and extraversion in a psychiatric population. In C. Barber (Ed.), *Evoked Potentials: Proceedings of an International Evoked Potential Symposium* (505–514). Lancaster: Medical and Technical Publications.

Polich, J. (1986). Normal variation of P300 from auditory stimuli. *Electroencephalography and Clinical Neurophysiology, 65,* 236–240.

Polich, J. (1989). Habituation of P300 from auditory stimuli. *Psychobiology, 17,* 19–28.

Polich, J., & Martin, S. (1992). P300, cognitive capability, and personality: A correlational study of University undergraduates. *Personality and Individual Differences, 13,* 533–543.

Pritchard, W. G. (1989). P300 and EPQ/STPI personality traits. *Personality and Individual Differences, IV,* 15–24.

Pritchard, W. S. (1991). The link between smoking and P: A serotonergic hypotheses. *Personality and Individual Differences, 12,* 1187–1204.

Raichle, M. E., Grubb, R. L., Gado, M. H., Eichling, J. O., & Ter-Pogonian, M. M. (1976). Correlation between regional cerebral blood flow and oxidative metabolism. *Archives of Neurology, 33,* 523–526.

Rawlings, D., & Borge, A. (1987). Personality and hemisphere function: Two experiments using the didactic listening technique. *Personality and Individual Differences, 8,* 483–488.

Rawlings, D., & Claridge, G. (1984). Schizotypy and hemisphere function: III. Performance asymmetries on tasks of letter recognition and local-global processing. *Personality and Individual Differences, 5,* 657–663.

Richards, M., & Eves, F. (1991). Personality, temperament and the cardiac defense response. *Personality and Individual Differences, 12,* 999–1004.

Risberg, J. (1980). Regional cerebral blood flow measurements by $133X_{e-}$ inhalation: Methodology and applications in neuropsychology and psychiatry. *Brain and Language, 9,* 9–34.

Risberg, J. (1986). Regional cerebral blood flow. In J. Hannay (Ed.), *Experimental Techniques in Human Neuropsychology* (514–543). Oxford: Oxford University Press.

Robinson, D. L. (1982). Properties of the diffuse thalamocortical system of human personality: A direct test of Pavlovian/Eysenckian theory. *Personality and Individual Differences, 3,* 1–16.

Robinson, D. L. (1986). On the biological determination of personality structure. *Personality and Individual Differences, 7,* 435–438.

Robinson, T. N., & Zahn, T. P. (1985). Psychoticism and arousal: Possible evidence for a linkage of P and psychopathy. *Personality and Individual Differences, 6,* 47–66.

Roger, D., & Jamieson, J. (1988). Individual differences in delayed heart-rate recovery following stress: The role of extraversion, neuroticism and emotional control. *Personality and Individual Differences, 9,* 721–726.

Royce, J. R., & Powell, A. (1983). *Theory of personality and individual differences: Factor, system, and processes.* Englewood Cliffs, N.J.: Prentice-Hall.

Saltz, E. (1970). Manifest anxiety: Have we misread the data? *Psychological Review, 77,* 508–573.

Saxton, P. M., Siegel, J., & Lukas, J. H. (1987). Visual evoked potential augmenting/reducing slopes in cats: 2. Correlations with behavior. *Personality and Individual Differences, 8,* 511–519.

Schalling, D., Asberg, M., Edman, G. (1984). *Personality and CSF monoamine metabolites.* Quoted by Zuckerman, 1991.

Schalling, D., Edman, G., & Asberg, M. (1987). Impulsive cognitive style and inability to tolerate boredom: Psychobiological studies of temperament and vulnerability. In M. Zuckerman (Ed.), *Biological bases of sensation seeking, impulsivity and anxiety.* Hillsdale, N.J.: Erlbaum.

Schalling, D., Edman, G., Asberg, M., & Oreland, L. (1988). Platelet MAO activity associated with impulsivity and aggressivity. *Personality and Individual Differences,* 597–605.

Shagass, C. (1954). The sedation threshold. A method for estimating tension in psychiatric patients. *Encephalography and Clinical Neurophysiology, 6,* 221–233.

Shigehisa, T., & Symons, J. R. (1973). Effect of intensity of visual stimulation on auditory sensitivity in relation to personality. *British Journal of Psychology, 64,* 205–213.

Shigehisa, P. M. J., Shigehisa, T., & Symons, J. R. (1973). Effects of intensity of auditory stimulation on photopic visual sensitivity in relation to personality. *Japanese Psychological Research, 15,* 164–174.

Sokolov, E. N. (1963). *Perception and the conditioned reflex.* Oxford: Pergamon.

Spence, K. W. (1964). Anxiety (drive) level and performance in eyelid conditioning. *Psychological Bulletin, 61,* 129–139.

Stassen, H. H., Lykken, D. T., & Bomben, G. (1988). The within-pair EEG similarity of twins reared apart. *European Archives of Psychiatry and Neurobiological Science, 237,* 244–252.

Stassen, H. N., Lykken, D. T., Propping, P., & Bomben, G. (1988). Genetic determination of the human EEG: Survey of recent results on twins reared together and apart. *Human Genetics, 80,* 165–176.

Stelmack, R. M. (1981). The psychophysiology of extraversion and neuroticism. In H. J. Eysenck (Ed.), *A Model for Personality* 38–64.

Stelmack, R. M., (1985a). Extraversion and auditory evoked potentials: Some empirical and theoretical considerations. In D. Papakostopoulos, S. Balter, and I. Martin (Eds.), *Experimental and Clinical Neurophysiology, 15,* 238–255.

Stelmack, R. M. (1985b). Personality and motor activity: A psychological perspective. In B. Kirkcaldy (Ed.), *Individual differences in movement* (153–213). Lancaster: Medical and Technical Publications.

Stelmack, R. M. (1990). Biological bases of extraversion: Psychophysiological evidence. *Journal of Personality, 58,* 293–311.

Stelmack, R. M., Achorn, E., & Michaud, A. (1977). Extraversion and individual differences in auditory evoked response. *Psychophysiology, 14,* 368–374.

Stelmack, R. M., & Geen, R. G. (1992). The psychophysiology of extraversion. In A. Gale and M. W. Eysenck (Eds.), *Handbook of Individual Differences: Biological Perspectives.* New York: Wiley.

Stelmack, R. M., Michaud, A., & Achorn, E. (1985). Extraversion, attention and habituation of the auditory evoked response. *Journal of Research in Personality, 19,* 416–428.

Stelmack, R. M., & Wilson, K. G. (1982). Extraversion and the effects of frequency and intensity on the auditory brainstem evoked response. *Personality and Individual Differences, 3,* 373–380.

Stemmler, G., Meinhardt, E. (1990). Personality, situations and physiological arousability. *Personality and Individual Differences, 11,* 293–308.

Stemmler, G., & Meinhardt, E. (1991). Personality, situations and physiological arousability: A reply to Zuckerman. *Personality and Individual Differences, 12,* 507–509.

Stenberg, G. (1992). Personality and the EEG: arousal and emotional arousability. *Personality and Individual Differences, 10,* 1094–1113.

Stenberg, G., Risberg, J., Warkentin, S., & Rosen, I. (1990). Regional patterns of cortical blood flow distinguishing extraverts from introverts. *Personality and Individual Differences, 11,* 663–673.

Stenberg, G., Rosen, I., & Risberg, J. (1988). Personality and augmenting/reducing in visual and auditory evoked potentials. *Personality and Individual Differences, 9,* 571–579.

Stenberg, G., Rosen, I., & Risberg, J. (1990). Attention and personality in augmenting/reducing of visual evoked potentials. *Personality and Individual Differences, 11,* 1243–1254.

Strelau, J., & Eysenck, H. J. (1987). *Personality dimensions and arousal.* New York: Plenum Press.

Strelau, J., Farley, F. H., & Gale, A. (1985). *The Biological basis of personality and behaviour,* (2 vols.). London: McGraw-Hill.

Szelenberger, W. (1983). Brainstem auditory evoked potentials and personality. *Biological Psychiatry, 18,* 157–174.

Thayer, R. E. (1970). Activation states as assessed by verbal response and four psychophysiological variables. *Psychophysiobiology, 7,* 86–94.

Thayer, R. E. (1989). *The Biopsychology of Mood and Arousal.* Oxford: Oxford University Press.

Thayer, R. E., Takahashi, P., & Pauli, J. (1988). Multidimensional arousal states, diurnal rhythms, cognitive and social processes, and extraversion. *Personality and Individual Differences, 9,* 15–24.

Theron, P. A. (1948). Peripheral vasomotor reactions as indices of basic emotional tension and lability. *Psychosomatic Medicine, 10,* 335–346.

Tranel, V. (1962). Effect of perceptual isolation on introverts and extraverts. *Journal of Psychiatric Research, 1,* 185–192.

Trouton, D. S., & Eysenck, H. J. (1961). Psychological effects of drugs. In H. J. Eysenck (Ed.), *Handbook of Abnormal Psychology.* New York: Basic Books.

Tucker, D. M., & Dawson, S. L. (1984). Asymmetric EEG changes as method actors generated emotions. *Biological Psychology, 19,* 63–75.

van der Merwe, A. B. (1948). The diagnostic value of peripheral vasomotor reactions in the psychoneuroses. *Psychosomatic Medicine, 10,* 347–354.

van der Merwe, A. B., & Theron, P. A. (1947). A new method of measuring emotional stability. *Journal of Genetic Psychology, 37,* 109–116.

Venables, P. (1963). The relationship between level of skin potential and fusion of paired light flashes in schizophrenic and normal subjects. *Journal of Psychosomatic Research, 1,* 279–287.

Venables, P. (1980). Primary dysfunction and cortical lateralization in schizophrenics. In M. Kruckow, D. Lebusann, & J. Angst, (Eds.), *Functional States of the Brain: Their Determinants.* Amsterdam: Elsevier.

Wenger, M. A. (1942). The stability of measurement of autonomic balance. *Psychosomatic Medicine, 4,* 94–95.

Wenger, M. A. (1948). Studies of autonomic balance in Army Air Force personnel. *Comparative Psychology Monographs, 19,* 1–111.

Wenger, M. A. (1957). Pattern analyses of autonomic variables during rest. *Psychological Medicine, 19,* 240–244.

Werne, P. F., Favery, H. A., & Janssen, R. (1973). Contingent negative variation and personality. *Electroencephalography and Clinical Neurophysiology, 37,* 739.

Werre, P. F. (1983). Contingent negative variation and interindividual differences. In R. Sinz & M. K. Rosenweig (Eds.), *Psychophysiology, memory, motivation and event-related potentials in mental operations* (pp. 337–342). Amsterdam: Elsevier.

Werre, P. F. (1986). Contingent negative variation: Relation to personality, and modification by stimulation of sedation. In J. Strelau, F. Farley, & A. Gale (Eds.), *The biological basis of personality and behavior, Vol. 2* (pp. 77–90). London: Hemisphere.

Wilson, G. D. (1981). Personality and social behavior. In H. J. Eysenck (Ed.), *A Model for Personality.* New York: Springer Verlag.

Wilson, G. D. (1990). Personality, time of day and arousal. *Personality and Individual Differences, 11,* 153–168.

Wright, R. A., Contrada, R. J., & Glass, D. C. (1985). Psychophysiological correlates of Type A behavior. In E. S. Rabbin & S. B. Manuck (Eds.), *Advances in behavioral medicine* (pp. 39–88). London: JAI Press.

Wundt, W. M. (1874). *Grundzuge der Physiologischen Psychologie.* Leipzig: Engelmann.

Zuckerman, M. (1991). *Psychobiology of personality.* Cambridge: Cambridge University Press.

Zuckerman, M., Murtaugh, T., & Siegel, J. (1974). Sensation-seeking and cortical augmenting and reducing. In H. J. Eysenck (Ed.), *The Measurement of Personality* (79–86). Lancaster: Medical and Technical Publishers.

Zuckerman, M., Simons, R. E., & Como, P. (1988). Sensation-seeking and stimulus intensity as modulators of cortical, cardiovascular, and electrodermal response: A cross-modality study. *Personality and Individual Differences, 9,* 361–372.

8

Cognitive Heterogeneity in Psychopathology
The Case of Schizophrenia

Gerald Goldstein

I. INTRODUCTION TO THE PROBLEM

In this chapter, I would like to address a problem in the psychology of individual differences that is particularly salient in the area of psychopathology. Since its beginnings, the field of psychopathology has been preoccupied with the problem of classification. What are the different forms of mental disorder, how can they be identified, and how can they be distinguished from one another? Classification, or diagnosis, is widely thought to be the essential basis for rational treatment. Psychiatry has coped with the matter essentially since its beginnings, and the history of psychiatric diagnosis is a long and complex one. Currently, two comprehensive classification systems are available, one contained in the four versions of the *Diagnostic and Statistical Manual of Mental Disorders* (DSM-I, II, III, and III-R) of the American Psychiatric Association and the other in the mental disorders section of the World Health Organization-sponsored *International Classification of Diseases* (ICD). The ICD has been periodically revised. DSM-III-R (APA, 1987) relates its diagnostic system to the ninth version (ICD-9-CM) (World Health Organization, 1978). DSM-IV is currently in preparation.

Without going into any detail about the history of these systems and the differences between them, it can be said that they are both largely descriptive in nature, and based upon clinical phenomenology. The term *clinical phenomenology* refers to the appearance of the mental disorders to an informed clinician, who makes the diagnosis largely on the basis of observation and interview. Reliance on laboratory tests is minimal, but laboratory tests may be used to rule out various medical disorders that may produce the clinical phenomenology of a mental disorder. The reliance on clinical description is based to a great extent on the fact that since the etiologies of most of the mental disorders are unknown, there are no definitive objective tests to diagnose or classify them. In that way, they differ from many of the disorders studied by neuropsychologists, for which there are objective neurological criteria available for diagnosis and classification.

The problem to which I alluded is that of heterogeneity within a single mental disorder. It occurs when individuals who fully meet the diagnostic criteria for the disorder vary extensively in ways that may be thought to be related to the disorder, and also in ways that are often seen in other disorders. Extensive heterogeneity of this type may, in the extreme case, call into question the construct validity of the disorder, but may also suggest the presence of an overarching disorder with various presentations. These presentations are often characterized as subtypes, implying that although there are individual differences, all of the subtypes belong to the same disorder and are not, in fact, different disorders.

Those knowledgeable about schizophrenia will no doubt be familiar with these matters. While there appears to be a general consensus that schizophrenia is a true diagnostic entity, there is also widespread acceptance that its presentation may vary greatly. This dual belief has led to the search for subtypes so that some order could be made out of the heterogeneity. Many of these subtypes (e.g., paranoid vs. nonparanoid) have also been developed on the basis of clinical phenomenology, and will not be dealt with in extensive detail here. DSM-III-R classifies schizophrenia into catatonic, disorganized, paranoid, undifferentiated, and residual types. Historically, four types were initially suggested by Kraepelin (1925) and E. Bleuler (1952); simple, catatonic, hebephrenic, and paranoid. Later, efforts were made to establish process vs. reactive (Becker, 1956; Herron, 1962), and good premorbid vs. poor premorbid types (Zigler, Levine, & Zigler, 1976). The very commonly used distinction between acute and chronic schizophrenia has been abandoned in DSM-III. To be diagnosed as schizophrenia the disturbance must persist for at least 6 months. Otherwise, the diagnosis of "brief reactive psychosis" or "schizophreniform disorder" is made.

More recently, attempts were made to establish typologies based upon various neurobiological considerations. Included here would be Crow's distinction between Type I and Type II schizophrenia (Crow, Cross, Johnstone, & Owen, 1984); Silverstein and Zerwic's (1985) distinction between neuropsychologically impaired and intact schizophrenics, and the attempt to establish a specific subtype characterized by a different course, and perhaps etiology, from other forms of schizophrenia; the so-called Kraepelinian type

(Keefe et al., 1987). All of these efforts appeared to maintain the assumption that schizophrenia is a single disorder or disease, and what was being described was a classificatory system within the framework of that disorder or disease.

It should be emphasized that this research regarding description, classification, and clinical phenomenology has been accompanied by a massive and broad-based effort to discover the etiology of schizophrenia. While the answer hasn't been found as yet, there have been numerous important findings. There is now little belief that schizophrenia is entirely a psychosocially acquired disorder, although various models involving environmental stresses triggering vulnerable individuals into schizophrenia have been suggested. There is a widely held belief that schizophrenia is a neurobiological disorder, probably of a developmental nature, and probably involving some form of neurochemical imbalance. Put more succinctly, it is commonly believed that schizophrenia is a form of brain dysfunction or disease (Henn & Nasrallah, 1982). Whether or not that condition is genetic remains controversial, but the existence of biological roots for the disorder is now a relatively uncontroversial issue.

For some time, there has been an intense interest in cognitive function in schizophrenia, and schizophrenia has been characterized as a cognitive disorder, as opposed to a mood or anxiety disorder. That is not to say that schizophrenics do not have mood- or anxiety-related difficulties, but there is a commonly held belief that abnormal thinking is at the core of the disorder. From the earliest days of scientific research with schizophrenic patients there has been a strong interest in studying the characteristics of their thinking, speech, and language, as contrasted with normal controls, or with patients with other disorders. Thinking has typically been described by such terms as concrete, idiosyncratic, autistic, and disorganized, whereas speech and language have been described by such terms as bizarre, fragmented, incoherent, and perseverative. In this research, extensive use was made of sorting and problem-solving tests, and tasks involving various language skills such as word definition and usage, language comprehension, and the ability to modify interpretation of words in changing contexts. Deficiencies were noted in all areas of this type when schizophrenic patients were compared with controls (Goldstein, 1978).

Difficulties with these studies of cognitive function in schizophrenia can be summarized with three points. First, it began to appear that regardless of the task chosen, schizophrenics always did worse than controls. Chapman and Chapman (1973) have characterized this phenomenon as the "general deficit syndrome," and proposed various methodological solutions; mainly the method of matched tasks. This method involves matching tasks for general difficulty level, and determining whether or not the patient group under study demonstrates selective impairment. Second, deficits noted in schizophrenics were also noted in patients with other disorders, notably brain-damaged patients (Watson, 1971; Watson, Thomas, Andersen, & Felling, 1968). That is, the cognitive deficits found lacked specificity. Third, the deficits found in samples of schizophrenics were not found in all patients, but only varying percentages of them. Thus, the deficits were also lacking in sensitivity.

In more recent times, the use of individual cognitive tasks has been supplemented by the application of extensive cognitive and neuropsychological test batteries. It thus became possible to generate sophisticated profiles of cognitive abilities, since the neuropsychological batteries typically assess a variety of functional domains including abstract reasoning and problem solving, attention, linguistic skills, memory, perceptual skills in vision, hearing, and touch, motor skills, and spatial abilities. It also became possible to contrast the functioning of the two sides of the body, and to generate test profiles that are characteristic for patients with lesions in various portions of the brain. Thus, schizophrenia has been characterized as a disorder involving the left temporal lobe (Flor-Henry, 1990) or the dorsolateral surface of the frontal lobe (Goldberg & Seidman, 1991). As a result of this research, it is now possible to characterize the status of schizophrenic patients in regard to each of the domains, and to offer hypotheses concerning neuroanatomic localization of dysfunction. When a full battery has been given to an individual patient, it is also possible to contrast one domain with another.

Nevertheless, the same problems persisted since there was once again overlap, particularly with brain-damaged patients, and extensive heterogeneity among the schizophrenic patients. These findings led to two major criticisms of neuropsychological tests. First, that these tests did not accurately discriminate between brain-damaged and schizophrenic patients. Second, that they were not helpful in the identification of schizophrenia, because of the great heterogeneity found among schizophrenic patients on these batteries. In summary, what we have in the case of schizophrenia, which would reasonably be characterized as a neuropsychological or neurobehavioral disorder by most contemporary authorities, is essentially a complete absence of a neuropsychological profile of the type found in many other neurobehavioral disorders. This excessive heterogeneity has raised questions ranging from issues related to the construct validity of the disorder itself to those related to the utility of neuropsychological assessment in the evaluation of schizophrenic patients. For this reason, it would seem to be profitable to adopt the theme of this volume, and investigate the nature of individual differences among schizophrenic patients. In what follows, we will limit our discussion to cognitive or neuropsychological heterogeneity, but heterogeneity has also been found in other areas as well. For example, some researchers have suggested that schizophrenia is a core disorder, with heterogeneity produced by differences in personality and experience. With regard to that view, however, we would suggest that the problem of heterogeneity in schizophrenia is not resolved by the readily agreed to consideration that schizophrenics, like all people, are unique individuals because of differences in personality and experience.

A. Demographic Sources of Heterogeneity

We will begin our discussion with the relatively simple and obvious sources of heterogeneity found on cognitive and neuropsychological tests. In normal individuals, heterogeneous results are found in samples of subjects who vary in age (Heaton, Grant, & Matthews, 1986; Reed & Reitan, 1963) and educational level (Finlayson, Johnson, & Reitan, 1977). Often, test scores are adjusted for age, education, or both, or there are age and education norms (Heaton, Grant, & Matthews, 1991). Some test scores are relatively more highly associated with age than with education, and the reverse is true for other tests. Within neuropsychology, the tests relatively more sensitive to age have been described as measuring "brain age" (Reitan, 1973) or "biological age" (Halstead & Rennick, 1962).

It has been suggested that in certain neurobehavioral disorders, the brain ages more rapidly than normal. Most of the research in this area has been done in alcoholism, in studies of the so-called premature aging hypothesis (Noonberg, Goldstein, & Page, 1985; Ryan & Butters, 1980), but obviously the brain also ages or deteriorates abnormally rapidly in the progressive dementias of the elderly. In the case of schizophrenia, there are two possibilities. One is that a source of heterogeneity is produced by age differences within the ranges typically produced by healthy normal individuals. Thus, if one tests a sample of schizophrenic patients, and if this sample consists of subjects that vary extensively in age, then variability in test scores will be found based upon that consideration alone. This possibility becomes more likely when the sample consists largely of late middle-aged to elderly subjects, where age-related changes on age-sensitive tests may be relatively large.

The other possibility is that schizophrenia is a disorder in which premature aging occurs. This possibility has been extensively studied, and raises crucial issues concerning the nature and course of schizophrenia. If it is the case, then heterogeneity will be potentiated because of the unusually large differences on age-sensitive measures that would occur between young and old schizophrenics. A consideration of this matter requires some digression from the specific matter of heterogeneity to a discussion of the nature and course of schizophrenia. The original formulation of the schizophrenic syndrome developed by Kraepelin (1925) indicated that schizophrenia was a degenerative dementia. The disorder was originally called *dementia praecox* in consideration of that belief. This view persisted for many years, but has been questioned recently. The more modern formulation of the course of schizophrenia is that it is an episodic disorder marked by an initial *first break,* followed by a period of substantial reduction in symptomatology, which may be followed by varying numbers of relapses marked by the return of symptoms of severity sufficient to require rehospitalization. There appear to be many reasons for the original, and now largely rejected, belief that schizophrenia is a degenerative disorder, including misdiagnosis of patients with structural neurological disorders, and an observation bias on the part of hospital staff who only saw relapsing patients and not those that remained in the community (M. Bleuler, 1978).

Research into schizophrenia and aging generally confirms the nondegenerative view. However, some investigators claim to have identified a subtype of schizophrenia, called the Kraepelinian type, that has a particularly poor prognosis and does, in fact, appear to be progressively degenerative. Most of the neuropsychological literature on schizophrenia and aging, regardless of whether a cross-sectional or longitudinal design was used, indicates that although schizophrenics typically do less well than nonschizophrenic controls on a variety of tests, age differences or changes are of the same order of magnitude in schizophrenics as they are in nonschizophrenics (Heaton & Drexler, 1987). That is, schizophrenics generally do not demonstrate premature aging. From the standpoint of the heterogeneity matter, it may be inferred that although age differences make for a source of heterogeneity in schizophrenia, they are not a source of greater heterogeneity than would be the case for nonschizophrenic individuals.

The matter of education has not been studied as thoroughly as age. Recently, it has been pointed out that there may be a difficulty in matching schizophrenic samples with controls for years of education, because the impact of a given number of years of education may be quite different for the schizophrenic or preschizophrenic than it would be for controls (Resnick, 1992). It may be more reasonable to match on the basis of socioeconomic status or performance on standard tests of educational achievement. However, it has been shown that years of education is a source of heterogeneity among schizophrenics, although it is not clear whether or not it is a more or less powerful source than is the case of nonschizophrenics. That is, although education may have a different impact on schizophrenics than from controls, we nevertheless find the same sometimes rather robust correlations between years of education and test scores in schizophrenics as we do in nonschizophrenics. An important consideration here is the cumulative effect of age and education. Typically, the neuropsychological test performance of older and less well-educated individuals may be quite disparate from performance levels found for younger and better educated individuals.

Other generally widely studied demographic variables, notably gender and ethnicity, have not been studied in schizophrenia with regard to their impact on heterogeneity in neuropsychological test performance. In general, it may be noted that while there appear to be important differences in brain organization between males and females, these differences typically need to be identified with specially designed experimental procedures, and there are very few gender differences on the more standard, commonly used neuropsychological tests. Race is typically confounded with educational and socioeconomic considerations, and is difficult to evaluate independently. Bilingualism is an important consideration in neuropsychological assessment, but we know of no studies of bilingual schizophrenics.

B. Iatrogenic Sources of Heterogeneity

The extensive use of psychoactive medication with schizophrenic patients has necessitated the study of the influence of these medications on test performance. This is the case because it is now clear that some, if not all, of these medications have cognitive effects. The practice of studying the cognitive effects of these medications in normal controls and animal models, although of great value in many applications, does not provide a solution to the problem of evaluating their effects in schizophrenic patients. That is the case for many reasons. There is a drug–disorder interaction commonly present, such that the cognitive effect of a drug on a normal individual may not be at all the same as its effect on a schizophrenic patient. Second, many patients take these medications over long periods of time, but their chronic effects are not readily amenable to investigation in normal controls. As a substitute for withdrawing patients from medication for research purposes, some investigators have suggested the method of studying misdiagnosed cases; individuals who were misdiagnosed as schizophrenic, medicated for lengthy periods of time, and then were withdrawn from medication when the misdiagnosis was discovered. However, this method does not deal adequately with the problem of drug–disorder interaction. It is now thought that the best available methodology is to study patients without a history of long-term usage while they are on and off medication, ideally within the framework of a randomized, counterbalanced, placebo, double-blind design.

In a comprehensive review of the effects of neuroleptic treatment on cognitive function in schizophrenia (Medalia, Gold, & Merriam, 1988) the complexity of this area is made apparent. Medication effects found may depend on the design of the study, the specific medication used—particularly with regard to anticholinergic effects or the anticholinergic effects of anti-Parkinsonism drugs that may be simultaneously administered—the way in which medication level is determined (dose vs. serum level), whether acute or chronic effects are being investigated, and the specific cognitive functions under examination. Furthermore, the literature is quite controversial, with a pattern of contradictory findings. Medalia et al. (1988) conclude that neuroleptic medication may adversely influence memory, fine motor coordination, and planning ability in schizophrenic patients with no adverse effects, or even improvement, in other areas of neuropsychological function. However, other authorities might not agree with even these modest conclusions. In our own research (G. Goldstein, 1990) we did not find medication status to be a robust contributor to heterogeneity. In general, it would appear that medication may influence heterogeneity in limited areas, but even there, that influence would not appear to be of major consequence.

Another possible iatrogenic source of heterogeneity is length of hospitalization. Some years ago, the practice of keeping schizophrenic patients hospitalized for lengthy periods of time was severely criticized, and indeed, that criticism was heeded. Among the numerous adverse effects attributed to long-term hospitalization was excessive cognitive deterioration. It was felt that the routinized, nondemanding nature of institutional life discouraged the exercise of cognitive abilities, thereby accelerating their deterioration. The limited literature on this topic has been largely negative. G. Goldstein and Halperin (1977) found significant neuropsychological differences between schizophrenics with more or less than 1 yr of hospitalization. However, studies by Johnstone, Owens, Gold, & Crow (1981) and by Harrow, Marengo, Pogue-Geile, and Pawelski (1987) found no association between length of hospitalization and cognitive function. In a more recent study, G. Goldstein, Zubin, and Pogue-Geile (1991) pointed out that the deterioration seen with increasing hospitalization among schizophrenic patients was essentially attributable to the anticipated changes associated with increasing chronological age.

C. Neurobiological Sources of Heterogeneity

There has been some suggestion that although schizophrenia may be a unitary disorder, the underlying biological etiological factors may not be the same in all cases. Perhaps the most well-known view is that of Crow and collaborators who presented the proposal that there are two syndromes of schizophrenia, one having a neurochemical and the other an infectious etiology (see Torrey & Peterson, 1976). Although the dopamine (DA) theory of schizophrenia is currently widely accepted as a viable working hypothesis, some investigators have implicated other neurotransmitters in the etiology of schizophrenia, notably norepinephrine (NE) and serotonin (SE). Thus, we have noradrenergic and serotonergic hypotheses of schizophrenia. To the best of my knowledge, it has not been seriously proposed that a typology exists here, and that there are DA schizophrenics and SE schizophrenics. However, Van Kammen (1991) has provided extensive evidence to suggest that the changes in the biochemical milieu, notably in the interaction between DA and NE, are related to state changes in individual patients, and are associated with relapse. It is therefore possible that neurochemical changes are largely associated with within-patient cognitive heterogeneity, because it is unlikely that the cognitive function of patients when actively psychotic will be the same as when they are clinically stable. The study of this area is made difficult by the fact that schizophrenic patients are frequently untestable while in actively psychotic episodes, and only become testable after attaining clinical stability, frequently associated with use of medication. While it is generally possible to test patients on and off medication, it is substantially more difficult to test patients in and out of psychotic episodes. Nevertheless, it would appear that neurobiological state–trait considerations would provide a considerable source of neuropsychological heterogeneity. Furthermore, the idea that there is more than one neurobiological etiology for schizophrenia should perhaps not be abandoned because of the existence of so-called refractory or treatment-resistant schizophrenics. These patients do not respond well to the standard neuroleptics, but some of them do respond well to other medications, possibly suggesting neurochemical differences from treatment respondents that are important for the etiology of schizophrenia in their cases.

Since the discovery of neuroradiological abnormalities in some young schizophrenic patients, indicated initially by computed tomography (CT) scan and more recently by nuclear magnetic resonance (MRI) imaging, a new neurobiological source of heterogeneity has been suggested. The Crow group (Crow et al., 1982) initially proposed this view, postulating that there were two types of schizophrenia. Each type was characterized on the basis of symptoms, type of illness, response to neuroleptics, outcome, presence of intellectual impairment, and postulated pathological process. Briefly summarizing this material, Type I schizophrenia is characterized by positive symptoms (delusions, hallucinations), acute schizophrenic illness, good response to neuroleptics, reversible outcome, absence of intellectual impairment, and the possibility that increased DA receptors is the etiology (the DA hypothesis). Type II schizophrenia is described as including primarily negative (deficit) symptoms, a chronic illness state, poor response to neuroleptics, a questionably irreversible outcome, intellectual impairment is sometimes present, and the proposed etiology is structural cell loss. Evidence is presented that the cell loss may have a viral etiology.

Initially, it was assumed that it was the Type II schizophrenics that had the abnormal CT scans or MRIs, reflecting the abnormal neuronal depletion ostensibly producing their illness. However, a recent study of identical twins discordant for schizophrenia called this view into question. It was shown that when the MRI scans of these twins were matched, it was clear by visual inspection that 12 of the 15 schizophrenic members of each pair had more brain atrophy than the unaffected twin (Suddath, Christison, Torrey, Casanova, & Weinberger, 1990). Many of the schizophrenic twin pairs' scans may have been read as normal using conventional clinical criteria, but relative to their cotwins they always showed greater neuronal depletion. The suggestion appears to be that all schizophrenics have abnormal neuronal depletion, but the effects may be too subtle to be interpreted as clinically abnormal in all cases. This study has been criticized because the schizophrenic index-twins had probably been medicated, and, being schizophrenic, may have lived under life circumstances that could be associated with the greater degree of atrophy (e.g., malnutrition, poor health maintenance). Nevertheless, it now appears simplistic to suggest that schizophrenics can be meaningfully classified on the basis of having a normal or abnormal CT scan or MRI.

What seems clearer is that there appears to be a relatively robust degree of association between quantitative measures of brain atrophy and quantitative neuropsychological test results. This extensive literature has been reviewed elsewhere, and will not be reviewed again here. A summary of these studies will be found in Walker, Lucas, and Lewine (1992) who conclude that "there appears to be a relationship between structural abnormalities and neuropsychological deficits" (p. 320). From the current perspective, that conclusion may be rephrased to suggest that variations in structural abnormalities of the brain may well be an important source of heterogeneity. However, several cautionary notes should be added to this conclusion. First, there is no clear evidence that structural brain abnormalities seen in schizophrenic patients have any causal relationship to the disorder. Secondly, this kind of relationship between neuropsychological function and structural integrity of the brain has been noted in several disorders, and is not at all unique to schizophrenia.

D. Comorbidity as a Source of Heterogeneity

Having schizophrenia does not exempt one from acquiring other disorders and diseases that afflict humans. Indeed, studies of longevity in schizophrenia (Allebeck, 1989) suggest that it is a life-shortening disease. One does not die of schizophrenia, but schizophrenics may be at slightly higher risk than nonschizophrenics for acquisition of a number of life-threatening diseases. Comorbidity of schizophrenia with substance abuse is commonly observed clinically (Alterman, 1985; Alterman, Erdlen, & Murphy, 1981). Furthermore, the life-style of many schizophrenics is likely to engender morbidity and mortality associated with such factors as trauma, malnutrition, and infection. This matter may become increasingly problematic in the case of the substantial number of schizophrenics who are homeless.

For some years we have been contrasting schizophrenic subjects with and without associated neurological dysfunction. Associated neuropsychological dysfunction was broadly defined, ranging from diagnosed frank neurological disease or disorder, frequent alcohol dementia, to isolated abnormal neurological findings, such as an abnormal electroencephalogram (EEG). In an earlier study utilizing the Halstead-Reitan battery (Reitan & Wolfson, 1985), a discriminant function analysis could not accurately classify schizophrenic subjects into cases with and without these associated disorders. However, in a later study, we found that associated neurological morbidity was a substantial contributor to cognitive heterogeneity (G. Goldstein, 1990).

At one time, it was thought that the neuropsychological deficits found in schizophrenic patients were largely attributable to inadequate case documentation, and that what was being observed were the consequences of undiagnosed brain disorder or disease, separate from the schizophrenia. Indeed, much of the early research in this area is substantially lacking with regard to adequate case documentation. To add to the difficulty, before the reforms of

DSM-III, the diagnosis of schizophrenia itself had some potential for unreliability (Gurland, 1974). Since the brain-imaging studies mentioned above, and improved practices in case documentation both on the neurological and psychiatric sides, it now appears that the neuropsychological deficits found in schizophrenics are genuine, and not the consequences of other illnesses. Indeed, in some cases these deficits can be quantitatively associated with measures of neuronal loss. However, we are left with the problem of reaching a more detailed understanding of the compounding of schizophrenia with other disorders that schizophrenics commonly acquire, notably alcoholism and malnutrition. These associated disorders continue to be a potential source of heterogeneity, but the delineation of what is attributable to the associated disorder and what is attributable to the schizophrenia is an admittedly difficult matter.

II. PROPOSED SOLUTIONS TO THE PROBLEM

A. A General Approach

It is apparent that the problem of heterogeneity cannot be productively investigated by traditional experimental approaches in which well-diagnosed schizophrenic patients, without neurological or other psychiatric illnesses, are contrasted with nonschizophrenic controls. The presence of statistical significance in these studies indicates that the average schizophrenic performs differently on the procedure under study than does the average nonschizophrenic. That is not the question we are asking. Put simply, we are inquiring about how many of the schizophrenics were different from controls on the procedure and how many were not. Generally, if one examines distributions of statistically significant findings, one will find areas of overlap in which members of the experimental group performed within the range of the control group. That does not compromise the value of the experiment, but one nevertheless has the option of looking at the average subject or the subjects within this overlapping range. Thus, when we ask questions such as, Are *all* schizophrenics impaired on the Wisconsin Card Sorting Test (WCST), or on a continuous performance task?, the answer is generally no. There typically are subjects with normal performance levels, or performance levels within the overlap range. Examination of data in this manner is generally within the purview of the psychology of individual differences, and requires borrowing from the methodology associated with that area. We are then necessarily led into the province of classification (e.g., schizophrenics who pass and who fail the WCST). Considering the material reviewed above, it is apparent that a single variable classification system of that type is not likely to be heuristic. Even the most simple natural phenomena would appear to require classification on the basis of more than one consideration.

Classification may be accomplished on the basis of rules or numerically. Aldenderfer and Blashfield (1984) provide an example of an ancient, rule-based method of classification of animals used by the Chinese, which we will quote in part. "Animals are divided into (a) those that belong to the Emperor, (b) embalmed ones, (c) those that are trained, (d) suckling pigs, (e) mermaids, etc." (p. 7). Rule-based classification is sometimes structured sequentially, as in the case of the taxonomic keys used to classify animals or plants. *DSM-III-R* does not use a formal taxonomic key for the establishment of diagnoses, but uses a system akin to what is called a polythetic key, in which not all of the rules must be met to make the classification. Thus, the first major rule for establishing the diagnosis of schizophrenia is presence of characteristic psychotic symptoms. Three subrules follow, but in the first of them only two of five stated conditions must be met: delusions, prominent hallucinations, incoherence, catatonic behavior, and flat or inappropriate affect. In the field of clinical neuropsychology, a more formal polythetic key for classifying types of brain damage was written by Russell, Neuringer, and Goldstein (1970).

The difficulty with the rule-based method as applied to the problem of cognitive heterogeneity in schizophrenia is that the rules are essentially unknown. While the types of schizophrenia contained in *DSM-III-R* are certainly useful clinical descriptions, neuropsychological dysfunction may be found in all of them. For example, G. Goldstein and Halperin (1977) found no difference on the Halstead-Reitan battery between paranoid and nonparanoid schizophrenics. In any event, while a rule-based method can be devised to classify phenomena, that classification only constitutes a convention, as in the example given above of how the ancient Chinese classified animals, without any basis in evidence suggesting that the classification represents some underlying reality. For example, the classification system proposed by Russell et al. (1970) would have been of very limited value were it not reasonably well validated against direct neurological evidence. Indeed, as this direct evidence accumulated, it was found that a component of the key was found to be less valid than originally reported, and thereby lost substantial value as a system for classifying brain-damaged patients (G. Goldstein & Shelly, 1982).

An alternative to rule-based classification is numerical taxonomy. The concept, developed by Sokal and Sneath (1963), involves the gathering of data on organisms of interest, estimating similarity among these organisms, and using a clustering method to assign these organisms to subgroups. These subgroups could then be studied at length to determine whether or not they represent different biological species, or, as we have put it, have some underlying reality. Ideally, the data to be gathered should be diverse, objective, and quantifiable. This concept of Sokal and Sneath, along with developments in statistical methods and computer technology, has led to a marked increase in the application of quantitative, empirical, classification methods. We propose that the application of these methods will be helpful in furthering our understanding of heterogeneity in schizophrenia. In this regard, our "role model" is the extensive work done with empirical classification in the field of learning disability. Like schizophrenia, learning disability is a cognitively heterogeneous condition, and numerous investigators have done extensive work with regard to applying empirical classification methods in the service of developing meaningful and clinically useful typologies (Rourke, 1985, 1991). The methods available can be divided into cluster analysis and factor analysis variants. In our own work, we have used cluster analysis, and so will provide a brief introduction to that method here. The reader familiar with cluster analysis may skip the next section.

B. A Brief Introduction to Cluster Analysis

Cluster analysis is a mathematical method of classifying cases on the basis of similarity. The concept of similarity is mathematically complex, but the most commonly used types of similarity indices are correlation coefficients and distance measures. As in other aspects of cluster analysis, the user has the choice of several methods, and may wish to use more than one of them to contrast solutions. There are also many clustering methods, each of which involves different ways of forming groups. In the biological and social sciences, the most commonly used methods are called the hierarchical agglomerative and iterative partitioning methods. The former methods sequentially merge similar cases, and can be represented visually by a tree diagram or dendogram. The iterative partitioning methods begin with a first partition of the data, followed by an iterative process involving change in cluster assignment, which is continued until no data points change clusters.

Morris, Blashfield, and Satz (1981) have provided a useful outline of steps needed to be taken for cluster-analytic work. First, subjects, variables, clustering method, and similarity method must be chosen. When the analysis is completed, the number of clusters must be determined and validation procedures should be accomplished. A distinction is made between internal and external validation. Internal validation has to do with the stability of the cluster solution itself. A stable and adequate solution is obtained when the clusters are distinct from each other, and when a similar solution is found using different clustering methods. If there is overlap in the space created by the different clusters or an excessive amount of reassignment of cases across clustering methods, then we do not have a stable or adequate solution. External validity involves the relationship between cluster membership and variables not included in the cluster analysis. Its establishment is often the most important part of cluster-analytic work, because cluster analysis can generate a highly internally valid solution that does not relate to any pertinent external variable. Indeed, artificial data sets generated with Monte Carlo procedures can generate clusters (Aldenderfer & Blashfield, 1984). Establishment of number of clusters is not a precise science. It is generally accomplished by visual inspection of the dendogram, sometimes supported by a form of *scree test* described by Aldenderfer and Blashfield that indicates when adding clusters contributes no new information.

We should add the usual cautionary notes concerning cluster analysis, several of which have been alluded to. Essentially, differing clustering methods can provide radically differing solutions, and determination of the number of clusters remains a reasonably subjective procedure. The user is encouraged to attempt more than one clustering method on the same data set and to cross-tabulate the various solutions. One might also wish to use caution in asserting that there is a given number of clusters, and to consider the varying implications of differing numbers of clusters.

In neuropsychological or psychometric cluster-analysis work, mean profiles for each cluster are often drawn, following appropriate transformation to standard scores. Thus, for example, if the cluster analysis was based on one of the Wechsler intelligence scales, the subtest profiles for each cluster would be drawn and contrasted. In these contrasts, however, the use of conventional significance levels among cluster-statistical comparisons is inappropriate, because significance has been essentially "forced" by the clustering method.

C. Classification Studies

In this section, we will try to integrate several studies in which cluster analysis was used to approach the problem of neuropsychological heterogeneity in schizophrenia. The first study (G. Goldstein & Shelly, 1987) involved cluster analyses of the Halstead-Reitan Battery, the Luria-Nebraska Neuropsychological Battery, and the Wechsler Adult Intelligence Scale (WAIS), each of which were applied to the same sample of heterogeneous psychiatric patients, 38% of whom were schizophrenic. In a related study of the utility of cluster analysis in classification of mixed neuropsychiatric patients (Schear, 1987), two cluster analyses of the WAIS and a modified Halstead-Reitan battery were performed with samples of 300 heterogeneous psychiatric patients. About 25% of the subjects were schizophrenic. Despite differences in clustering methods and, to some extent, tests, the two studies arrived at remarkably similar results. First, both studies achieved stable, internally reliable clustering solutions. Second, both studies found no relationship between psychiatric diagnosis and cluster membership. This latter finding was also noted in a study by Townes et al. (1985) using a different classification method. For purposes of the present discussion, it is noted that there were schizophrenic subjects in each subgroup in each study. It would appear that the G. Goldstein and Shelly (1987) and Schear (1987) studies, with backup from Townes et al. (1985) have impressively demonstrated the heterogeneous performance of schizophrenic patients on the most commonly used standard neuropsychological tests. Put another way, the results indicate the lack of diagnostic specificity of these tests.

The results of these studies can be seen as leading to an important, decisive point concerning further investigation. One could draw the conclusion that it is time to abandon the use of standard neuropsychological tests for assessment of schizophrenics because of the tests' lack of specificity. The lack of sensitivity has also been documented because of the finding that some schizophrenic patients perform normally on these tests. Another formulation would be the one adopted by Townes et al. (1985), who asserted that neuropsychological tests assess a different dimension of behavior from psychiatric diagnosis. That dimension may be called "competence" or some similar term reflecting adaptive abilities. Thus, while neuropsychological tests may not identify schizophrenia, they may asses the schizophrenic's pattern and level of competence.

An alternative strategy, and the one we adopted, is to admit that cognitive function is indeed heterogeneous in schizophrenia, but despite the fact that there is no single sensitive and specific schizophrenic profile, it is nevertheless of interest to explore the heterogeneity within schizophrenia. The reason that it is interesting is because schizophrenics with different cognitive profiles may have varying characteristics associated with those profiles. For examples, using the Crow group's descriptive categories, they may have varying symptom patterns, type of illness, response to neuroleptics, outcome, and postulated pathological process. Thus, the tests could potentially provide pertinent information about the patient that would not be obtainable from an interview based on clinical phenomenology, or from laboratory procedures. With regard to laboratory procedures, however, there may be meaningful associations among them and the neuropsychological tests.

A very preliminary effort to adopt that approach was taken in the final classification study to be discussed (G. Goldstein, 1990). This study only evaluated schizophrenic patients, and concentrated on one domain of cognitive function: abstraction and problem-solving ability. There is abundant evidence accumulated over many years and in many settings that impairment of abstract reasoning is a core feature of schizophrenia. From the early clinical studies of K. Goldstein (1959) and Hanfmann and Kasanin (1942) with sorting tests, to the most recent investigations with the WCST given in association with sophisticated functional-imaging procedures (Goldberg & Weinberger, 1988; Weinberger, Berman, & Zec, 1986), the practice of giving sorting tests to schizophrenics and noting their frequent failures has persisted. Being interested in heterogeneity, we therefore asked the question, are all schizophrenics impaired on abstraction and problem-solving tasks?

Abstraction and problem solving have several components. K. Goldstein and Scheerer (1941) provide a list of modes of behavior that reflect the abstract attitude along with appropriate assessment tasks. We have borrowed from this list, but sometimes changed the task. The ability to discover a rule or principle that organizes diverse material is generally assessed with sorting tests, as is the ability to modify rules as required by circumstances. The

Halstead Category Test (HCT) is a good test for conceptual rule learning, while the WCST places a greater emphasis on rule modification, or shifting sets. Planning ahead ideationally is also a component of the abstract attitude, and is assessed well with the Halstead Tactual Performance Test (HTPT), which is performed most efficiently when some systematic plan or strategy is utilized. Another aspect of abstraction is the ability to keep various aspects of a stimulus or problem in mind simultaneously. Part B of the Trail Making Test (TRB) is a good measure of this ability, because the subject must keep track of the number and letter sequence at the same time during the course of performing the task. Thus, we administered a battery consisting of the HCT and HTPT, the WCST, and the TRB. Only data from Part B of the TRB were considered.

Data from 136 male schizophrenic patients who received all of these tests were cluster analyzed, using a hierarchical agglomerative method (Ward's method), and squared Euclidean distance as the similarity measure. A five-cluster solution was proposed. Internal validity of the solution was evaluated by plotting the clusters in discriminant function space and inspecting for overlap. The five clusters appeared as clear separate entities, with minimal overlap. Thus, heterogeneity among our schizophrenic subjects was clearly found, but could be accounted for on a preliminary basis within the framework of five empirically derived subtypes.

In what follows, I would like to expand on the description of these subtypes beyond what could be covered in the original paper. The emphasis will be on external validity. Following that, implications for future study will be discussed. The test score means and standard deviations for each cluster are summarized in Table 1.

Cluster 1   This cluster contained 34 subjects (25% of the sample). The mean age was 39 yr (SD = 10.6), with a mean of 12.6 yr of education (SD = 2.8), and a mean WAIS Full Scale IQ of 99.6 (SD = 13.3). On the abstraction tests,

TABLE 1
Results for Abstraction and Problem-Solving Tests by Cluster

| Test | Cluster | | | | | | | | | |
| | 1 | | 2 | | 3 | | 4 | | 5 | |
	M	SD	M	SD	M	SD	M	SD	M	SD
Category (errors)	71.8	27.0	37.2	14.6	109.7	23.0	91.9	25.9	86.6	17.9
TPT[a] (minutes)	21.0	6.1	15.7	3.9	27.4	3.8	23.5	6.5	20.4	6.4
TRB (seconds)	118.4	15.6	64.0	12.4	296.4	10.7	181.9	18.5	81.2	2 0.5
WCST—categories	3.2	2.2	3.4	2.2	1.0	1.2	3.2	2.4	2.2	1.9
WCST—cards	120.9	15.4	117.9	18.4	128.0	.0	118.0	17.3	127.4	3.1

[a]Abbreviations: TPT, Tactual Performance Test; TRB, Trail Making Test; WCST, Wisconsin Card Sorting Test.

this cluster obtained mean scores on the HCT, TPT, and TRB that would be rated as mildly impaired by the Russell et al. (1970) norms. These ratings are not available for the WCST, but it is noted that the average subject did not complete all six categories. According to the Heaton (1980) manual, normals between the ages of 40 and 49 achieved a mean of 4.8 categories (*SD* = 1.8), but Heaton's normative sample had a somewhat higher mean IQ (112.4) than did ours. The clinical information available on this cluster indicates that 62% of its members were receiving antipsychotic medication, and 42% of them had definite neurological dysfunction (e.g., alcohol dementia or head trauma) in addition to schizophrenia.

From the standpoint of cognitive function, this cluster, while of average general intelligence, has mild deficits in abstraction and problem solving. However, in that regard they have a relatively flat profile, with no apparent discrepancies in their capacities to find rules, shift sets, develop strategies, or exercise simultaneous function. We would speculate that the cluster represents what is probably the most typical cognitive profile for chronic schizophrenia, in which all conceptual abilities are mildly, but uniformly impaired. Within the Crow group system, we would think that the modal case would be a Type II schizophrenic, because of the presence of substantial intellectual impairment. It is also noteworthy that only 62% of the cluster was taking antipsychotic medication, the lowest percentage of all five clusters. This finding at least raises the question of whether they were poor neuroleptic responders, but we don't have data with which to answer that question.

Cluster 2   This cluster is composed of 28 cases (20.6% of the sample). The mean for age was 30.9 yr (*SD* = 8.9), 13.5 yr for education (*SD* = 2.2), and the mean IQ was 108.6 (*SD* = 11.0). The Russell et al. (1970) ratings would classify the scores for the HCT, TPT, and TRB as normal. However, the cluster did only slightly better than Cluster 1 on the WCST. Heaton's manual indicates that his subjects under age 40 achieved a mean of 5.6 categories. Seventy-five percent of this cluster was taking antipsychotic medication, and 29% had definite neurological comorbidity.

This cluster is of particular interest because, with the exception of the WCST, they performed within the average range on the abstraction and problem-solving tests. We therefore have a subgroup of schizophrenics that appears to have little in the way of impaired abstraction and problem-solving ability. Unfortunately, we were not able to score for perseverative errors and other qualitative indicators now available for the WCST, and so cannot provide detailed information concerning our subjects' performances. However, from the categories-achieved score relative to the good score obtained on the HCT, it would appear that these subjects may have a specific deficit in conceptual shifting, but not in other aspects of abstract reasoning. Recently, we have seen this same pattern in individuals with autism (Minshew, Goldstein, Muenz, & Payton, 1992).

In consideration of the relatively high percentage of patients taking anti-psychotic medication and the relative absence of intellectual impairment, we would speculate that this cluster largely reflects the Crow group's Type I schizophrenia. These patients should have a high proportion of positive symptoms, and are likely to be good neuroleptic responders. They may be less "chronic" than the patients in Cluster 1.

Cluster 3 There were 26 subjects in Cluster 3 (19.1% of the sample). The means for age were 43.8 (*SD* = 9.4), 9.8 for years of education (*SD* = 3.6), and 78.9 for full-scale IQ (*SD* = 9.1). On the abstraction tests, the Russell et al. ratings would be severely impaired for the HCT, moderately impaired for the TPT, and very severely impaired for TRB. The mean for the WCST was only one category achieved. Ninety-one percent of the subjects in this cluster were receiving antipsychotic medication, and 50% had definite neurological comorbidity.

This cluster obviously reflects significantly impaired function, and substantially more impairment than was found for the other clusters. The mean age is slightly higher than was the case for the other clusters, but is not in a range in which one would expect to see any but the earliest manifestations of the presenile dementias. Nevertheless, these test scores would be consistent with significant dementia. One might speculate that the cluster represents the severe end of the Crow group Type II subtype, but it seems more likely, given the severity of the impairment, that many of the patients in the cluster would meet criteria for Kraepelinian schizophrenia. Kraepelinian schizophrenics have been described as severely deteriorated and showing more negative symptoms than most schizophrenics. Admittedly, the average educational level for Cluster 3 is lower than is the case for the other clusters, but even for the mean educational level of 9.8 yr obtained by the cluster, a mean IQ of 78.9 is likely to reflect at least moderate general deterioration.

Cluster 4 There were 22 subjects in this cluster (16.2% of the sample). The mean age was 40.5 (*SD* = 9.4); the mean educational level was 11.4 yr (*SD* = 2.8), and the mean IQ was 94.0 (*SD* = 13.5). On the abstraction tests, the HCT, TPT, and TRB means would all be rated as moderately impaired. WCST performance was comparable to Clusters 1 and 2. Ninety-one percent of the subjects in this cluster were taking antipsychotic medication, and 55% of them had definite neurological comorbidity.

This cluster is much like Cluster 1, both of which exhibit flat profiles. However, this group is somewhat more impaired. More of them were taking antipsychotic medication than was the case for Cluster 1, and there was more neurological comorbidity. Therefore, they may be a more impaired group generally, but we see no evidence for a qualitative difference from Cluster 1.

Cluster 5 Cluster 5 contains 26 subjects (19.1% of the sample). The mean age is 38.6 yr (SD = 13.0), with a mean education level of 11.8 yr (SD = 2.30, and a mean IQ of 96.0 (SD = 10.1). The HCT would be rated as moderately impaired, the TPT as mildly impaired, and TRB as average. WCST performance was poor relative to Clusters 1, 2, and 4. Ninety-one percent of the cases were taking antipsychotic medication, and there was 39% definite neurological comorbidity.

The most interesting feature of this cluster would appear to be the relative preservation of psychomotor speed. That is, they do relatively well on the TPT and TRB, both of which are scored on the basis of time to completion. It is noteworthy that with the exception of Cluster 2, Cluster 5 has the lowest prevalence of neurological comorbidity. In the original paper, we commented on the strong association between the TRB and neurological comorbidity. Thus, Cluster 5 might not differ qualitatively from 1 and 4 with regard to schizophrenia-related parameters, but may be relatively less impaired on the basis of structural brain damage not associated with schizophrenia itself.

To summarize these admittedly speculative considerations, it would appear that the pattern of heterogeneity obtained divides cases into what would really appear to be three subgroups; one with generalized moderate impairment, one with relative absence of impairment and functioning that approaches normal, and one with dense impairment suggestive of significant deterioration. A fourth group resembles the first subgroup, but is somewhat more severely impaired. A fifth group also resembles the first subgroup, but has superior psychomotor speed capabilities, and a lower prevalence of neurological comorbidity. Thus, the clustering seems to have been based upon a combination of schizophrenia-related and neurological comorbidity-related considerations.

Some note should be made of the contributions of the various tests to the clustering. The WCST contributed little differentiation, perhaps because the categories-achieved score was not sufficiently refined. Perseverative errors and similar refined scores should be used in future research. With regard to the WCST, the data presented might imply that none of the subjects achieved all six categories. That was not the case. About 20% of the subjects achieved all six categories but they were distributed into all of the clusters, except cluster 3. Further evidence for heterogeneity in WCST performance ranging from severe impairment to intact function has been reported by Braff et al. (1991) and by Condray, Steinhauer, and Goldstein (1992). The TPT also did not make a great contribution, although it did highlight the generally superior performance of Cluster 2. The HCT showed clear differences among some, but not all of the clusters. The test that contributed most to cluster assignment was the TPT. In the discriminant function analysis used to plot cluster membership, it was the first-entered variable. We have already pointed to its probable association with neurological comorbidity.

III. CONCLUSIONS BASED ON HETEROGENEITY STUDIES

We will summarize these findings in terms of identifying sources of heterogeneity. Age is a source of heterogeneity, but, with regard to cognitive function, it does not appear to be a more influential source in schizophrenia than it is in normal aging. Educational status also does not appear to be a major source of heterogeneity in schizophrenics in particular. Length of hospitalization does not appear to be a source of heterogeneity separate from what is attributable to chronological age. Medication status does not appear to be a substantial source of heterogeneity as far as we have been able to determine. In the study in which it was considered, medication status did not vary substantially among clusters. However, medication may play some role in state-related heterogeneity, particularly anticholinergic drugs. Neurological comorbidity is a substantial source of heterogeneity, but is far from being a sole determinant. As an illustration, there was a higher prevalence of neurological comorbidity in Cluster 4 than in Cluster 3, yet Cluster 3 was substantially more impaired. The prevalence of neurological comorbidity in the G. Goldstein (1990) study sample as a whole may seem surprisingly high, but the specific diagnoses were mainly congenital and developmental disorders, alcohol dementia, and head trauma, all of which are reasonably common among chronic schizophrenics of early middle age. It is noted that none of the subjects in the studies reviewed in which neurological comorbidity was considered had acute neurological disease. Perhaps one of the most important findings is that the cognitive heterogeneity found, while influenced more or less by all of these considerations, is not entirely accounted for by them. We have speculated that current classification schemes, notably those proposed by the Crow group and the isolation of Kraepelinian schizophrenia as a separate syndrome, may constitute additional, schizophrenia-related influences.

IV. CURRENT STATUS OF THE PROBLEM AND FUTURE DIRECTIONS

We have been proceeding on the basis of a distinction between heterogeneity associated with schizophrenia-related parameters and extra schizophrenia parameters that influence cognitive function. Thus far, more has been accomplished with the latter set of variables than with the former. In essence, we have learned that some extra schizophrenia parameters, notably age and neurological comorbidity, do influence heterogeneity, but don't tell the whole story. At one time, there was some belief that extra schizophrenia parameters were largely responsible for impaired neuropsychological test performance by schizophrenic patients. These findings were attributed to such considerations as recondite neurological disease, substance abuse, or medication effects. Elsewhere (G. Goldstein, 1991) we attempted to provide documentation that

the neuropsychological deficits found in schizophrenics are not epiphenom-
ena associated with schizophrenia-related symptomatology, such as impaired
attention, lack of motivation, idiosyncratic test-taking attitude, or failure to
cooperate. Summarizing briefly, this view was largely discredited by findings
involving robust correlations between neuropsychological deficits and neuroimaging data (Walker et al., 1992), as well as by activation studies in which
testing was done online with measures of cerebral metabolism (Goldberg &
Weinberger, 1988).

As we view the matter, the major problem is that of relating neuropsychological heterogeneity to schizophrenia-related parameters. We have suggested
the possibility that this additional heterogeneity can be understood in terms
of some of the proposed neurobiological subtypes of schizophrenia. Thus far,
we have speculated that a severe deficit in a range consistent with the presence
of dementia may appear in patients who would meet criteria for the Kraepelinian subtype. Patients with lesser degrees of impairment may fit into the
Crow group's Type II schizophrenia, whereas patients with little in the way
of cognitive deficit may be Crow group Type I patients. There admittedly are
ambiguities here, because it is possible that patients with mild to moderate
deficits might be evolving into Kraepelinian schizophrenia, while the relatively unimpaired patients may eventually acquire schizophrenia-related impairment. In essence, developmental considerations might be an aspect of
heterogeneity, with patients going from one status to another. The types
would not be permanent in that case, but would be developmental phenomena. We have already indicated that state–trait considerations may also have
this temporal characteristic. Only longitudinal studies will answer these
questions.

The pursuit of other sources of heterogeneity may be accomplished
through extensive refinements and expansions of the G. Goldstein (1990)
study. Basically, there need to be procedural refinements, and a great deal
more information has to be gathered concerning the patients. Procedural
matters include replication of the cluster analysis, application of additional
clustering methods with the original and replication samples, and diagnostic
refinement. The diagnoses of schizophrenia were made clinically and not
through the use of structured interviews designed to generate *DSM-III-R*
diagnoses. Indeed, much of the data were collected during the *DSM-II* era. It
is possible that a new sample of cases diagnosed with rigorous, state-of-the-art
procedures will show reduced heterogeneity. It would also be of interest to
study a subsample of schizophrenics with no identifiable neurological comorbidity associated with disorders other than schizophrenia.

Additional information needed would include clinical subtype diagnoses,
quantitative neuroimaging findings, response to medication, and course and
outcome data. Most pertinently here, our speculation that certain of the
clusters reflect the Crow subtypes or Kraepelinian schizophrenia need to be
objectively verified.

There are some data that suggest that heterogeneity will appear even among carefully diagnosed cases. Indeed, there is some evidence that it appears in well-diagnosed cases that are genetically related. It is reported that the Genain quadruplets (Rosenthal, 1963) exhibited heterogeneity on an attention task. Recently, Condray, Steinhauer, Van Kammen, and Zubin (1992) studied a set of monozygotic twins concordant for paranoid schizophrenia, who were diagnosed independently by two psychiatrists using structured psychiatric interviews. Neither twin had examinational evidence or a history of neurological comorbidity. Nevertheless, they showed striking differences on the WCST, TPT, and other neuropsychological tests. Thus, heterogeneity was found even in this exceptionally well-matched pair of subjects.

REFERENCES

Aldenderfer, M. S., & Blashfield, R. K. (1984). *Cluster analysis.* Beverly Hills, CA: Sage.

Allebeck, P. (1989). Schizophrenia: A life shortening disease. *Schizophrenia Bulletin, 15,* 81–89.

Alterman, A. I. (1985). Relationships between substance abuse and psychopathology: Overview. In A. I. Alterman (Ed.), *Substance abuse and psychopathology* (pp. 1–12). New York: Plenum.

Alterman, A. I., Erdlen, F. E., & Murphy, E. (1981). Alcohol abuse in the psychiatric population. *Addictive Behaviors, 6,* 69–73.

American Psychiatric Association (1987). *Diagnostic and statistical manual of mental disorders,* (3rd ed.). Washington, D.C.: Author.

Becker, W. (1956). A genetic approach to the interpretation and evaluation of the process-reactive distinction in schizophrenia. *Journal of Abnormal and Social Psychology, 53,* 229–236.

Bleuler, E. (1952). *Dementia praecox or the group of schizophrenias.* New York: International Universities Press.

Bleuler, M. (1978). *The schizophrenic disorders: Long-term patient and family studies.* Translated by S. M. Clemens. New Haven: Yale University Press.

Braff, D. L., Heaton, R., Kuck, J., Cullum, C., Moranville, J., Grant, I., & Zisook, S. (1991). The generalized pattern of neuropsychological deficits in outpatients with chronic schizophrenia with heterogeneous Wisconsin Card Sorting Test results. *Archives of General Psychiatry, 48,* 891–898.

Chapman, L. J., & Chapman, J. P. (1973). *Disordered thought in schizophrenia.* New York: Appleton-Century-Crofts.

Condray, R., Steinhauer, S., & Goldstein, G. (1992). Language comprehension in schizophrenics and their brothers. *Biological Psychiatry, 32,* 790–802.

Condray, R., Steinhauer, S., Van Kammen, D. P., & Zubin, J. (1992). Dissociation of neurocognitive deficits in a monozygotic twin pair concordant for schizophrenia. *Journal of Neuropsychiatry and Clinical Neuroscience, 4,* 449–453.

Crow, T. J., Cross, A. J., Johnstone, E. C., & Owen, F. (1982). Two syndromes in schizophrenia and their pathogenesis. In F. A. Henn & H. A. Nasrallah (Eds.), *Schizophrenia as a brain disease* (pp. 196–234). New York: Oxford University Press.

Finlayson, M. A. J., Johnson, K. A., & Reitan, R. M. (1977). Relationship of level of education to neuropsychological measures in brain-damaged and non-brain-damaged adults. *Journal of Consulting and Clinical Psychology, 45,* 536–542.

Flor-Henry, P. (1990). Neuropsychology and psychopathology: A progress report. *Neuropsychology Review, 1,* 103–123.

Goldberg, E., & Seidman, L. J. (1991). Higher cortical functions in normals and in schizophrenic: A selective review. In S. R. Steinhauer, J. H. Gruzelier & J. Zubin (Eds.), *Handbook of schizophrenia: Vol. 5: Neuropsychology, psychophysiology and information processing* (pp. 553–597). London: Elsevier.

Goldberg, T. E., & Weinberger, D. R. (1988). Probing prefrontal function in schizophrenia with neuropsychological paradigms. *Schizophrenia Bulletin, 14*, 179–183.

Goldstein, G. (1978). Cognitive and perceptual differences between schizophrenics and organics. *Schizophrenia Bulletin, 4*, 160–185.

Goldstein, G. (1990). Neuropsychological heterogeneity in schizophrenia: A consideration of abstraction and problem solving abilities. *Archives of Clinical Neuropsychology, 5*, 251.264.

Goldstein, G. (1991). Comprehensive neuropsychological test batteries and research in schizophrenia. In S. R. Steinhauer, J. H. Gruzelier, & J. Zubin (Eds.), *Handbook of schizophrenia, Vol. 5: Neuropsychology, psychophysiology and information processing* (pp. 525–551). London: Elsevier.

Goldstein, G., & Halperin, K. M. (1977). Neuropsychological differences among subtypes of schizophrenia. *Journal of Abnormal Psychology, 86*, 34–40.

Goldstein, G., & Shelly, C. (1982). A further attempt to cross-validate the Russell, Neuringer, and Goldstein neuropsychological keys. *Journal of Consulting and Clinical Psychology, 50*, 721–726.

Goldstein, G., & Shelly, C. (1987). The classification of neuropsychological deficit. *Journal of Psychopathology and Behavioral Assessment, 9*, 183–202.

Goldstein, G., Zubin, J., & Pogue-Geile, M. F. (1991). Hospitalization and the cognitive deficits of schizophrenia: The influences of age and education. *The Journal of Nervous and Mental Disease, 179*, 202–206.

Goldstein, K. (1959). Concerning the concreteness in schizophrenia. *Journal of Abnormal and Social Psychology, 59*, 146.

Goldstein, K., & Scheerer, M. (1941). Abstract and concrete behavior: An experimental study with special tests. *Psychological Monographs, 53*.

Gurland, B., & Professional Staff of the U.S.–U.K. Cross-National Project (1974). The diagnosis and psychopathology of schizophrenia in New York and London. *Schizophrenia Bulletin, 1*, 80–102.

Halstead, W. C., & Rennick, P. M. (1962). Toward a behavioral scale for biological age. In C. Tibbitts & W. Donahue (Eds.), *Social and psychological aspects of aging* (pp. 866–872). New York: Columbia University Press.

Hanfmann, E., & Kasanin, J. (1942). Conceptual thinking in schizophrenia. *Nervous and Mental Disease Monographs, 67*.

Harrow, M., Marengo, J., Pogue-Geile, M. F., & Pawelski, T. J. (1987). Schizophrenic deficits in intelligence and abstract thinking: Influence of aging and long-term institutionalization. In N. E. Miller & G. D. Cohen (Eds.), *Schizophrenia and aging: Schizophrenia, paranoia and schizophreniform disorders in later life* (pp. 133–144). New York: Guilford.

Heaton, R. K. (1980). *A manual for the Wisconsin Card Sorting Test.* Odessa, FL: Psychological Assessment Resources, Inc.

Heaton, R. K., & Drexler, M. (1987). Clinical neuropsychological findings in schizophrenia and aging. In N. E. Miller & G. D. Cohen (Eds.), *Schizophrenia and aging: Schizophrenia, paranoia and schizophreniform disorders in later life* (pp. 145–161). New York: The Guilford Press.

Heaton, R. K., Grant, I., & Matthews, C. G. (1986). Differences in neuropsychological test performance associated with age, education, and gender. In I. Grant & K. Adams (Eds.), *Neuropsychological assessment of neuropsychiatric disorders* (pp. 100–120). New York: Oxford University Press.

Heaton, R. K., Grant, I., & Matthews, C. G. (1991). *Comprehensive norms for an expanded Halstead-Reitan battery.* Odessa, FL: Psychological Assessment Resources, Inc.

Henn, F. A., & Nasrallah, H. A. (Eds.) (1982). *Schizophrenia as a brain disease.* New York: Oxford University Press.

Johnstone, D. G., Owens, C., Gold, A., & Crow, T. J. (1981). Institutionalization and the defects of schizophrenia. *British Journal of Psychiatry, 139*, 195–203.

Herron, W. G. (1962). The process-reactive classification of schizophrenia. *Psychological Bulletin, 59*, 329–343.

Keefe, R. S. E., Mohs, R. C., Losonczy, M. F., Davidson, M., Silverman, J. M., Kendler, K. S., Horvath, T. B., Nora, N., & Davis, K. L. (1987). Characteristics of very poor outcome schizophrenia. *American Journal of Psychiatry, 144*, 889–895.

Kraepelin, E. (1925). *Dementia praecox and paraphrenia.* Edinburgh: Livingston.

Medalia, A., Gold, J., & Merriam, A. (1988). The effects of neuroleptics on neuropsychological test results of schizophrenics. *Archives of Clinical Neuropsychology, 3*, 249–271.

Minshew, N. J., Goldstein, G., Muenz, L. R., & Payton, J. B. (1992). Neuropsychological functioning in nonmentally retarded autistic individuals. *Journal of Clinical and Experimental Neuropsychology, 14*, 749–761.

Morris, R., Blashfield, R., and Satz, P. (1981). Neuropsychology and cluster analysis: Potentials and problems. *Journal of Clinical Neuropsychology, 3*, 79–99.

Noonberg, A., Goldstein, G., & Page, H. A. (1985). Premature aging in alcoholism: "Accelerated aging" or "Increased vulnerability?" *Alcoholism: Clinical and Experimental Research, 9*, 449–469.

Reed, H. B. C., & Reitan, R. M. (1963). A comparison of the effects of the normal aging process with the effects of organic brain damage on adaptive abilities. *Journal of Gerontology, 18*, 177–179.

Reitan, R. M. (September, 1973). *Behavioral manifestations of impaired brain function in aging.* Paper presented at the annual meeting of the American Psychological Association, Montreal, Canada.

Reitan, R. M., & Wolfson, D. (1985). *The Halstead-Reitan neuropsychological battery: Theory and clinical interpretation.* Tucson: Neuropsychology Press.

Resnick, S. M. (1992). Matching for education in studies of schizophrenia. *Archives of General Psychiatry, 49*, 246.

Rosenthal, D. (Ed.) (1963). *The Genain quadruplets.* New York: Basic Books.

Rourke, B. P. (Ed.) (1985). *Neuropsychology of learning disabilities: Essentials of subtype analysis.* New York: Guilford.

Rourke, B. P. (Ed.) (1991). *Neuropsychological validation of learning disability subtypes.* New York: Guilford.

Russell, E. W., Neuringer, C., & Goldstein, G. (1970). *Assessment of brain damage: A neuropsychological key approach.* New York: Wiley-Interscience.

Ryan, C., & Butters, N. (1980). Learning and memory impairments in young and old alcoholics: Evidence for the premature-aging hypothesis. *Alcoholism: Clinical and Experimental Research, 4*, 288–293.

Schear, J. M. (1987). Utility of cluster analysis in classification of mixed neuropsychiatric patients. *Archives of Clinical Neuropsychology, 2*, 329–341.

Silverstein, M. L., & Zerwic, M. J. (1985). Clinical psychopathological symptoms in neuropsychologically impaired and intact schizophrenics. *Journal of Consulting and Clinical Psychology, 53*, 267–268.

Sokal, R., & Sneath, P. (1963). *Principles of numerical taxonomy.* San Francisco: W. H. Freeman.

Suddath, R. L., Christison, G. W., Torrey, E. F., Casanova, M. F., & Weinberger, D. R. (1990). Anatomical abnormalities in the brains of monozygotic twins discordant for schizophrenia. *The New England Journal of Medicine, 322*, 789–794.

Torrey, E. F., & Peterson, M. R. (1976). The viral hypothesis of schizophrenia. *Schizophrenia Bulletin, 2*, 136.

Townes, B. D., Martin, D. C., Nelson, D., Prosser, R., Pepping, M., Maxwell, J., Peel, J., & Preston, M. (1985). Neurobehavioral approach to classification of psychiatric patients using a competency model. *Journal of Consulting and Clinical Psychology, 53*, 33–42.

Van Kammen, D. P. (1991). The biochemical basis of relapse and drug response in schizophrenia: review and hypothesis. *Psychological Medicine, 21,* 881–895.

Walker, E., Lucas, M., & Lewine, R. (1992). Schizophrenic disorders. In A. E. Puente & R. J. McCaffrey (Eds.), *Handbook of neuropsychological assessment* (pp. 309–334). New York: Plenum.

Watson, C. G. (1971). Separation of brain damaged from schizophrenic patients by Reitan-Halstead pattern analysis. *Psychological Reports, 29,* 1343.

Watson, C. G., Thomas, R. W., Andersen, D., & Felling, J. (1968). Differentiation of organics from schizophrenics at two chronicity levels by use of the Reitan-Halstead organic test battery. *Journal of Consulting and Clinical Psychology, 32,* 679.

Weinberger, D., Berman, K., & Zec, R. (1986). Physiological dysfunction of dorsolateral prefrontal cortex in schizophrenia: I. Regional cerebral blood flow evidence. *Archives of General Psychiatry, 43,* 114.125.

World Health Organization (1978). *Mental disorders: Glossary and guide to their classification in accordance with the ninth revision of the international classification of diseases.* Geneva: World Health Organization.

Zigler, E., Levine, J., & Zigler, B. (1976). The relation between premorbid competence and paranoid-nonparanoid status in schizophrenia: A methodological and theoretical critique. *Psychological Bulletin, 83,* 303–313.

9

Neuropsychology of Temperament

Jose J. Gonzalez, George W. Hynd, and Roy P. Martin

I. INTRODUCTION

Over the past 15–20 yr, there has been an explosion of scientific interest in the description, measurement, antecedents, and sequelae of temperament assessed during childhood. This increase in interest can be traced directly to the research activity of Alexander Thomas, Stella Chess, and colleagues (Thomas & Chess, 1977; Thomas, Chess, & Birch, 1968). Collectively they demonstrated through the New York Longitudinal Study that social, emotional, and attentional characteristics of children measured very early in their life were predictive in important ways of later occurring behaviors, particularly in relation to social and emotional problems.

Based on the promise of these early studies, researchers from a variety of fields have examined temperament as a potential contributor to mother–child interactions (e.g., Crockenberg, 1986), peer group relationships (e.g., Attilli, 1990), educational achievement (e.g., Martin, 1989b), teacher interactions with exceptional children (e.g., Keogh, 1982), and hospitalizations (Huttunen & Nyman, 1982).

The concept of temperament has been most appealing to researchers in the areas of behavioral genetics, personality theory, and developmental psychology, who are collectively seeking the genetic foundations of temperament and the precursors to adult personality and psychopathology. Thus, most of the research has focused on the behavior of the child, its developmental stability, and behavioral sequelae. Conversely, very little research has focused exclusively on the neuropsychological basis of those behaviors characterized as temperamental in nature.

In this context, and with these considerations in mind, it is the purpose of this chapter to examine a relatively meager amount of literature on the neuropsychological basis of temperament and to discuss the potential relevance of research on related topics. However, prior to such a discussion, it seems important to briefly examine the boundaries of the concept of temperament as we see it, and some of the most important theoretical implications of the concept.

II. THE NATURE OF TEMPERAMENT

There are a variety of points of view on the definition of temperament (Bates, 1989; Goldsmith et al., 1987). There does, however, seem to be general agreement that temperament consists of individual differences in behavioral tendencies that are present early in life and are relatively stable across time and situations. Also, temperament is considered to be of biological origin with some theorists giving particular emphasis to genetic determinants (e.g., Buss & Plomin, 1975, 1984).

There is also general agreement that the set of behavioral characteristics that are considered to be temperamental is relatively small, ranging from three (Eysenck, 1959, 1970) to nine (Thomas & Chess, 1977) characteristics or traits. Table 1 provides a summary of these different perspectives. At present,

TABLE 1
Four Views of Temperament Structure

Thomas and Chess (1977)	Presley and Martin (in press)	Buss and Plomin (1984)	Eysenck & Eysenck (1975)
Activity Level[a]	Activity Level	Activity Level	Introversion-Extraversion
Approach–Withdrawal	Inhibition		Neuroticism
Adaptability	Adaptability	Sociability	
Emotional Intensity	Negative Emotionality		Psychoticism
Mood		Emotionality	
Persistence	Persistence		
Distractibility			
Threshold	Threshold		
Rhythmicity	Rhythmicity		

[a]Placement of temperament characteristics in adjacent columns is intended to be suggestive of similarity of constructs. For example, negative emotionality in the Presley and Martin list and mood in the Chess and Thomas list have something in common with Emotionality in the Buss and Plomin list. The reader should be aware that in almost every case, the Buss and Plomin and the Eysenck constructs are not highly similar to the constructs in the other two lists. Specifically, the Buss and Plomin construct of sociability indicates more the desire to be with others than social inhibition or social adaptability imply. The constructs of psychoticism and neuroticism take on much broader meaning than any of the temperament traits listed.

however, there is a growing consensus that seven dimensions of temperament account for most of the variance measured by current assessment devices (Martin, Wisenbaker, & Huttunen, in press; Presley & Martin, in press; Rothbart, 1989). These include activity level (the amount and vigor of gross and fine motor movement exhibited by the child); social inhibition (often referred to as approach–withdrawal; denotes the tendency to withdraw or inhibit activity in novel social situations); adaptability (the ease and speed of adjustment to an altered social environment, including family and school rules); negative emotionality (often referred to as negative mood or emotional intensity; denotes the frequency and vigor of expressions of anger, hostility, and related emotions); task persistence and distractibility (the tendency to continue difficult tasks without being distracted); rhythmicity (the temporal regularity of biological functions such as feeding, eating, elimination, and sleep); and threshold (level of stimulation necessary to produce a response in visual, auditory, or tactile modality) (Martin et al., in press).

The emphases given to heritability and stability and in the definition of temperament have led some to believe that temperament is established at birth, and is relatively impervious to the effects of the environment. Such a position is clearly an oversimplification and masks the complexity of how the nature of temperament is conceptualized. We endorse the position taken by Gottlieb (1991) and Cairns (1991) that individual development is best viewed from the systems perspective. In this perspective, individual development and indeed temperament are thought of as being hierarchically organized into multiple levels (gene, cell, organ, organ system, organism, behavior, environment) and that all levels are interactively influenced by the other levels.

Another important aspect of temperament theory is that stability of temperamental characteristics is thought to occur primarily because environmental factors that affect behavior are correlated with the genetic factors that affect the same behaviors. This occurs for several reasons. First, the behavioral tendencies of the child tend to foster specific environmental responses from others. For example, the agreeable or adaptive child may receive more nurturance than the emotionally negative child. Second, as individuals mature they seek environmental niches in which their individual differences are most adaptive and hence, most likely to be reinforced. In this context, one child in elementary school may seek activities in which social skills play an important role, whereas another withdraws from such activity. In both instances, the child's behavioral predisposition plays a role in creating the most reinforcing environment. In this way, the environment the child experiences strengthens the already existing genetic predisposition thus creating progressively more stability in temperament behavior as the child matures (Scarr & McCartney, 1983). Empirically, the stability of temperamental behavior has been found by most researchers to (1) increase from infancy into childhood, (2) to be affected by situational factors, and (3) to be generally in a range considered high moderate if the situation remains relatively constant and low moderate in altered circumstances. To illustrate the latter two points, Martin et al. (1986) demonstrated that six teacher-rated temperamental characteristics assessed over a 6-month interval in a test–retest design were correlated in a range of .69 to .85 (ratings were made by some teachers at both time intervals). Over a 12-month time period the correlations ranged from .43 to .67 for five of the six characteristics. For one characteristic the correlation was near zero. The second rating was made by a different teacher than the one who provided the first rating, and the rating was based on behavior observed in a different classroom context.

Thus, the concept of temperament as it is understood in contemporary literature is thought of as comprising a small number of traits that are present early in life, are biologically rooted, and are relatively stable. However, there is no assumption that these behavioral characteristics are unchangeable, or that the environment of the child does not have a critical role in shaping and reinforcing temperamental characteristics. In this context then, the study of the genetic contribution to the manifestation of temperament seems particularly relevant.

A. The Genetic Basis of Temperament

There is growing support for the hypothesis that temperament has a sizable heritability component, although the evidence for each of the characteristics is not uniform. Goldsmith and Alansky (1987) suggested that the evidence in support of heritability is strongest for the characteristics of negative emotionality and activity level, but moderate heritability estimates have been found for other characteristics.

One of the most persuasive studies is that of Braugart, Plomin, Defries, and Fulker (1992). They studied 161 1- and 2-yr-old children who were adopted at birth and 189 nonadoptive children of the same age who were matched on gender, number of siblings, and education and occupation level of father. These children were examined using the Bayley tests, and were rated on the Bayley Infant Behavior Record (IBR) immediately following the examination. Three factors were extracted from the IBR: affect–extraversion, activity, and task orientation. These three factors were the same as the first three factors isolated from the same measure by Matheny (1980). In addition to the adoptive and nonadoptive sample, 194 monozygotic (MZ) twins and 96 dizygotic (DZ) 1- and 2-yr-old twins were also rated on the IBR.

The findings of this study are as follows: First, intraclass correlations for MZ twins on the IBR were higher on all three factors than the DZ twins and the adoptive sibling pairs. The former correlations ranged from .40 and .50 for affect–extraversion; .30 and .58 for activity, and .50 and .58 for task orientation (first correlations were for 1-yr-old subjects, second for 2-yr-old subjects). Correlations for DZ twins were .05 and .03 for affect–extraversion,

.25 and .18 for activity, and .21 and .20 for task orientation. For nonadoptive sibling pairs the correlations were .05 and .20 (affect–extraversion), .20 and .21 (activity), and .23 and .22 (task orientation). The correlations for adoptive sibling pairs showed near-zero order correlations indicating a negligible impact of shared environment on infant temperament. Second, the average heritability for all three IBR factors was 44%. Affect–extraversion showed a 42% genetic component. Task orientation showed a 44% additive genetic component, and activity showed a 47% genetic component.

Based on the results of this study, and others that have obtained similar results (see reviews by Plomin, 1990b; Goldsmith et al., 1987), it can be estimated that from 40–50% of the variance found in questionnaire-measured temperament can be attributed to genetic factors (Plomin, 1990b). This is a very similar figure to that found for personality traits (Plomin, 1990a). This suggests that temperament may not be more heritable than many personality variables. One possible interpretation of these findings is that personality variables are so influenced by and reflective of temperamental characteristics that much of the heritability observed in adult personality is temperament based. Another interpretation is that temperamental characteristics seen early in life are not necessarily more genetically influenced than personality characteristics assessed in adolescence or adulthood. Certainly if temperament is to a certain degree heritable, then one might suspect that associated behaviors and cognitive abilities might also manifest differently according to temperamental traits.

III. NEUROLOGICAL FOUNDATIONS OF TEMPERAMENT

In this section, we concentrate on several contemporary theorists who have addressed possible connections between temperament and brain functions. First, the theories of Pavlov and Eysenck will be briefly described, as their work provided some of the conceptual foundation for the remaining theorists. In addition, the contemporary theories of Jeffrey Gray, Marvin Zuckerman, and Jan Strelau will be discussed as they provide some of the most specific ideas currently available on the physiological foundations of temperament. Finally, the ideas of Jerome Kagan will be presented. This work has a special place in theories of temperament because it focuses on one temperamental dimension (social inhibition). Kagan's work is not so much a grand theory of neurological foundations of temperament, but an extensive set of data against which to judge his own theory of social inhibition and the more general theories of others.

As will be evident, there is considerable overlap in the terminology and theoretical processes that are discussed by these theorists. Nonetheless, we will attempt to draw some comparisons and attempt to integrate common elements from these various theories.

A. Ivan Pavlov

Pavlov described two major properties of the nervous system—strength and mobility. Further, the balance of these processes was also given considerable weight in his theorizing. Strength of the nervous system is defined as the strength of the excitatory process and mobility is defined as the capacity to shift from one process to the other (excitatory to inhibitory). He maintained that from these properties could be delineated four types of nervous systems, which correspond to the classical four types of temperament as defined by Hippocrates and Galen. For example, strong and mobile nervous systems were equivalent to the sanguine types of temperament. Pavlov stated that the type of nervous system is innate and only slightly susceptible to environmental or rearing influences. According to Pavlov, temperament is the innate type of the nervous system or the typical behavioral traits of an individual that are conditioned by the innate type of his nervous system.

Based on seminal research in animals, Pavlov, Teplov, and Nebylitsyn investigated the notion of the innate nature of the basic human nervous system properties (strength, mobility, and balance of nervous processes) and attempted to distinguish the innate from the acquired. Teplov and Nebylitsyn (1963) found that there is a relatively stable relationship between sensitivity (Pavlov's nervous system strength) and efficiency. The weaker the stimulus that elicits a perceptible response (the higher the sensitivity) and the weaker the stimulus that starts to lower efficiency (the lower the resistance), the higher is an individual's reactivity; conversely, a low-reactive individual is characterized by low sensitivity and high resistance. As we will see, these concepts reemerge in Strelau's research.

B. Jan Strelau

Strelau's work is unique in contemporary temperament research because it is grounded in Eastern, particularly Russian, psychology, and can be traced directly back to the work of Pavlov. As many other theorists do, Strelau (1985) draws a fundamental distinction between temperament and personality. In Strelau's conceptualization, temperament is a basic property of the nervous system of both animals and humans, whereas personality is a product of external social conditions and is an essentially human phenomenon. He believes that dimensions of behavior, which have a major physiological component and are a result of biological evolution, are an aspect of temperament, whereas all the traits or mechanisms that are influenced by social–historical conditions are personality. This does not mean, of course, that temperamental traits are not affected by environmental influences.

According to Strelau, temperament is defined as a set of relatively stable traits that are observable in the *energy level* of behavior and in the temporal patterns of reaction. Such traits are seen as *formal* because they have no content and do not affect behavior directly; rather, temperament, as a regulative mechanism, is manifested in behavior indirectly. Strelau defines "energy level" as including all traits that make up individual differences and the corresponding physiological mechanisms involved in the accumulation and release of energy. This system is located at endocranial, autonomic, and brain stem levels (the reticular-activating system), including those cortical areas or systems that are linked with lower centers and are involved in the regulation of excitation. These physiological mechanisms are seen as operating as a system with stable characteristics and in which the contributions of elements vary between individuals.

Two basic temperamental features of energy level are reactivity and activity. Reactivity is a stable pattern of response to stimulus intensity in a person. It extends from extreme sensitivity, both sensory and emotional to resistance or efficacy. Activity is associated with what Strelau calls the behavioral energy level, which varies in the intensity or frequency in which different kinds of tasks are undertaken. For example, when attention-deficit hyperactivity disorder (ADHD) children or severely mentally retarded children are low on arousal (hypothetically) they may engage in behaviors (motor overactivity) so that they attain an optimal level of arousal activation. There is evidence that at a metabolic level this is exactly what happens when a hyperactive child receives stimulant medication. Lou, Henriksen, and Bruhn (1984) and Lou, Henriksen, Bruhn, Børner, and Nielsen (1989) found that in hyperactive children metabolic levels in the region of the basal ganglia were low compared to normals. When given stimulant medication, metabolic levels normalized and the motor overactivity ceased. Thus, there is some support that impulsivity and motor overactivity are indeed related to low levels of arousal and potentiation of dopamine (DA) reuptake (via stimulant medication) increases arousal to normal levels, which is in turn associated with motor inhibition. However, and of some interest theoretically, these effects seem most pronounced in children with "pure" ADHD because in children with co-occurring symptoms of aggression, the effects are less dramatic (Martier, Halperin, Sharma, Newcorn, & Sathaye, 1992).

With regards to Strelau's notion of reactivity, an extreme response might be considered as anxious behavior. One could postulate that the central nervous system (CNS) is overaroused in such a condition and indeed children who are diagnosed as having an anxiety disorder get worse when given stimulant medication (Swanson, Kinsbourne, Roberts, & Zucker, 1978).

Thus two features as articulated by Strelau, although seen as independent, interact in a complex manner both as a function of individual development and of situational requirements. Highly reactive people are those who are particularly sensitive to stimulation and have what Strelau refers to as a high stimulation-processing coefficient, which reflects a deficiency within physiological mechanisms that serve to suppress stimulation. Activity is related to the notion of optimal level of arousal in that the person acts to increase or decrease incoming stimulation to achieve his or her optimum level. Thus, weakly reactive people will seek additional stimulation and activation, whereas highly reactive people will act to reduce it. Different temperamental traits will react to the same environment differently, depending on the child's level of reactivity. For example, a highly reactive child will avoid situations and activities that bring along strong stimulation, whereas less reactive children seek highly stimulating activities.

In the course of development, temperamental traits start to reveal themselves in more and more complex and diverse reactions, inciting others to interact with the child in accordance to the type of his temperament. the temperament induces change in the environment, which in turn has certain consequences for the development of the individual's personality. This indirectly affects the formation of personality by establishing behavioral patterns that Strelau believes affect personality development.

C. Hans Eysenck

The writings of Hans Eysenck have played an important role in conceptualizations of personality structure, temperament, and the biological basis of personality. Many have given Eysenck credit for being the first modern personologist to view personality from a neuropsychological basis (e.g., Zuckerman, 1991).

Perhaps the most influential aspect of Eysenck's theory relates to the structure of personality. In 1947, Eysenck first articulated the basic theory that he has been explicating and reformulating since that time. This theory posited that there are three basic dimensions of personality: introversion-Extraversion (E), Neuroticism (N), and Psychoticism (P). Each of these elements have been considered as dispositional; that is, they are thought of as having significant heritability. Thus, Eysenck's theory could be thought of as a theory of temperament, although he does not use this term.

One of the difficulties in describing the nature of each of the three dimensions of this theory is that the instruments used to measure these dimensions, and the understanding of what each instrument assesses, has changed in important ways over the years. For example, the first measure of the dimensions of the model was the Maudsley Personality Inventory (Eysenck, 1959). This measure included only the E and N dimensions; the P dimension was not included in a self-report questionnaire until 1975. In the 1959 version, the E dimension was assessed using scales developed by Guilford. However, it was determined that the E and N dimensions were correlated and the theoretical model had postulated that these dimensions would be orthogonal. Therefore, in 1964 Eysenck and Eysenck published the Eysenck Personality Inventory in which the correlation between E and N was removed. The E dimension was composed of two types of items in the Eysenck Personality Inventory: those tapping sociability and impulsivity. A new scale was developed in 1975 (Eysenck & Eysenck, 1975), referred to as the Eysenck Personality Questionnaire, which included the P dimension. Because of the theoretical requirement to maintain orthogonal dimensions, impulsivity items tended to end up in the P dimension, and to be taken out of the E dimension. Thus, E became primarily a sociability factor.

In addition to changes in dimensional content, and relationship between dimensions, the N and P scales have become progressively more normalized (Zuckerman, 1991) over the years. For example, as the name psychoticism implies, this scale was devised originally as an index of a trait that was thought of as a precursor of many types of mental illness. The original scale included items tapping paranoid ideation, sadistic tendencies, and aggressiveness, among others. Over the years, many of these more pathological items have been eliminated.

At the current time, the narrow traits subsumed by each of the macrotraits of the model can be summarized as follows (Zuckerman, 1991, p. 13):

E: Sociable, lively, active, assertive, sensation seeking, carefree, dominant, surgent, venturesome
N: Anxious, depressed, guilt feelings, low self-esteem, tense, irrational, shy, moody, emotional
P: Aggressive, cold, egocentric, impersonal, impulsive, antisocial, unemphatic, creative, tough-minded

The theory that Eysenck developed in support of his dimensional theory has remained relatively unchanged over the past 35 yr, although there have been changes in some of the specific anatomical links as the field of neuropsychology has grown more sophisticated and the data testing the theory have accumulated. E is thought of as being controlled by the ascending reticulocortical-activating system. The function of this neuroanatomical system is said to be the control of the level of stimulation of the CNS. It is thought of as regulating the organism's response to stimulation received through the sensory channels and its associated cortical registration. As Zuckerman (1991) describes the process, "Depending on the load on the brain, the cortex may open channels for incoming stimulation or close them through descending pathways to the RAS [reticular-activating system]" (pg. 134). In this model, introverts are individuals characterized by high levels of CNS arousal, and for this reason, they engage in various types of behaviors to limit the amount of external stimulation they receive. For example, they engage in less social behavior or tend to avoid it because other people can be a source of strong stimulation. Extraverts, on the other hand, are characterized by a lower level of CNS arousal, and they engage in activities that increase stimulation, such as more social interaction.

The limbic system (referred to by Eysenck as the visceral brain) is the set of structures that determines the level of N that is exhibited and experienced by the individual. According to Eysenck, limbic reactivity that has a low threshold is thought to characterize the high scorer on the N dimension, and the reverse is thought to be true of the low scorer. The neuroanatomical systems controlling P functioning have been given far less attention by Eysenck than have the other dimensions.

This theory of personality, particularly the neuropsychological aspects of the theory, have been criticized for years for being simplistic. For example, as Zuckerman (1991) points out, the theory makes little distinction among the various systems of the limbic brain, with all simply being thought of as controlling emotionality. This level of nonspecificity is not acceptable in light of the more current work of Gray, for example. However, Eysenck's theory has proven so provocative that it has served as one of the major theories against which other researchers have reacted, and has been the starting point for much subsequent research.

D. Jeffrey Gray

Gray's (Gray, 1991) major hypothesis is that temperament reflects individual differences in predispositions towards particular kinds of emotion and that emotions are states of the CNS (central or conceptual nervous system) elicited by reinforcing agents. Much of Gray's work is derived from his own research with animals, although he discusses potential applications to humans.

Gray defines emotion as "consistency of states elicited by stimuli or events which have the capacity to serve as reinforcers for instrumental behavior," (p. 106). A reinforcer is then any stimulus that, if made contingent upon a response, alters the future probability of a reoccurrence of that response. His model describes three fundamental emotion systems located in the human and animal CNS, each of which (1) responds to a separate subset of reinforcing events with specific types of behavior, and (2) is mediated by a separate set of interacting brain structures that process specific types of information. The three systems are referred to as the behavioral inhibition system (BIS), the fight–flight system (F/FLS), and the behavioral approach system (BAS). Gray postulates that individual differences in the functioning of these three systems, and their interaction, underlie human temperament.

The BIS is the only one for which a corresponding human emotion, anxiety, can validly or reliably be identified. The eliciting stimuli involved in this system are conditioned stimuli associated with punishment, conditioned stimuli associated with the omission or termination of reward, or novel stimuli. The BIS is thought to act as a *comparator*, checking predicted against actual events, and being able to interrupt motor execution programs if they do not match. The system also modulates the control of exploratory behavior, by increasing arousal and attention. In the case of anxiety, it is assumed that when a mismatch between predicted and actual events occurs, the motor program is halted and outputs of the BIS attempt to take more information in (via heightened arousal and attention) and resolve the difficulty. The major neural structures in this system are found in the septohippocampal system (SHS). The core of the comparator function is attributed to the subicular area. The prefrontal cortex is allotted the role of providing the comparator with information concerning the current motor program. The monoaminergic pathways that ascend from the mesencephalon to innervate the SHS are believed to alert the entire system under conditions of threat and divert its attention to deal with the threat. Although the above analysis is speculative, research into information processing (cognitive function) and positron emission topography (PET) studies on humans support some of these hypotheses.

Gray's next system, the F/FLS has been described as having very tentative relationships with the rest of the model, and in turn is highly schematic in terms of neuroanatomical substrates. The F/FLS responds to unconditioned aversive stimuli and unconditioned defense aggression or escape behavior.

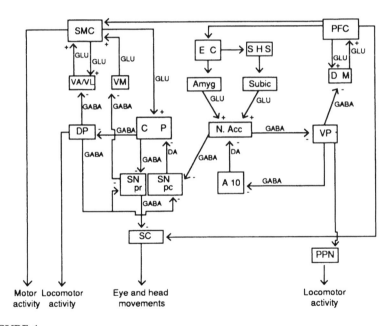

FIGURE 1

Major structures composing the neural basis of the Behavioral Approach System and its interrelations with the Behavioral Inhibition System. Structures: SMC = sensorimotor cortex; PFC = prefrontal cortex; EC = entorhinal cortex; SHS = septohippocampal system; Subic = subicular area; Amyg = amygdala; VA/VL = N. ventralis anterior and ventralis lateralis thalami; VM = N. ventralis medialis thalami; DM = N. dorsalis medialis thalami; DP = dorsal palladium; VP = ventral palladium; CP = caudate-putamen; N. Acc = N. accumbens; SNpr = substantia nigra, pars reticulata; SNpc = substantia nigra, pars compacta; A 10 = N. A 10 in ventral tegmental area; SC = superior colliculus; PPN = penduculopontine nucleus. Transmitters: GLU = glutamate; DA = dopamine; GABA = γ-aminobutyric acid. (Reprinted from Gray, 1991, p. 121)

Stimulation and lesion experiments have located three important levels of structure that appear to have F/FLS function(s): the amygdala, which inhibits the medial hypothalamus via the ventroamygdalofugal pathway, which in turn inhibits the final output pathway in the central gray via the dorsal longitudinal Bundle of Shütz. There has been little research on the information-processing (cognitive) activities of the F/FLS, and no direct links with a corresponding human emotion been found, but anger and terror have been postulated as potential behavioral manifestations controlled by this system.

The BAS is a simple positive feedback system, activated by stimuli associated with reward or with the termination or omission of punishment. It functions to increase spatiotemporal proximity to such stimuli. This system in general is capable of guiding the organism to the goals it needs for survival. The key components of this "motor programming system," as some researchers call it (Groves, 1983; Swerdlow & Koob, 1987), are the basal ganglia, the dopaminergic fibers that ascend from the mesencephalon to innervate the basal ganglia, thalamic nuclei closely linked to the basal ganglia, and neocortical areas closely linked to the basal ganglia. These components form two closely interrelated subsystems. The caudate motor system is a nonlimbic cortico-striato-pallido-thalamic-midbrain circuitry that encodes the specific content, in terms of relationships between stimuli, responses, and reinforcement, of successive steps in a goal-directed motor program. This system seems to be vitally important in the regulation of motor activity, particularly with regard to motor inhibition. The accumbens motor system is a limbic cortico-striato-pallido-thalamic-midbrain circuitry, which (1) switches between steps in the motor program and (2) interacts with the SHS, monitoring the routine operation of the motor program towards the intended goal. The SHS performs a monitoring function in which this feedback is received from the projection to nucleus accumbens from the subiculum. There has been little research relating the BAS to actual human emotion, but given the functions attributed to this system, Gray postulates the BAS underlies such states as pleasurable anticipation, elation, or happiness. (For a more detailed description of structures, pathways, and corresponding hardware of any of these systems see Gray, 1991.)

In Gray's writing (Gray, 1991), temperament and personality are used interchangeably as meaning what remains of individual differences once general intelligence and other specialized cognitive functions, such as visuospatial or verbal ability, have been removed. It is assumed that temperament reflects individual differences in predispositions towards particular kinds of emotion. Gray postulates that the variation of the emotion systems among individuals is what gives rise to personality. A dimension of personality corresponding to individual differences in the intensity of functioning of the BIS, F/FLS, and BAS at the high end of the continuum, may be high trait anxiety, high propensity for aggressive and defensive behavior, and behavior motivated by positive reinforcement and accompanying pleasurable emotions, respectively.

Going back to the example of ADHD children, using Gray's model, ADHD children would be deficient in their BIS (Canivez, 1992; Milich, 1993). This type of conceptualization resembles Satterfield's (1978) theory of hyperactivity, which proposes a low CNS arousal combined with an insufficient inhibitory control over behavior. In turn, response cost treatments, as well as stimulant medication, may work best with these types of children because they allow the children to inhibit their behavior to a more normal degree and therefore strengthen their BIS.

E. Marvin Zuckerman

Zuckerman's psychobiological theory of personality is strongly influenced by Eysenck's personality theory and Gray's theory of the neurological foundations of personality. In a concluding chapter of a recent book (Zuckerman, 1991), which reviews much of the extant literature on the biological (genetic, neuroanatomical, neurotransmitter, and pharmacological) basis of personality, Zuckerman posits a complex neurological model of personality. Unlike Gray, the model is formulated primarily in terms of the neurochemical mechanisms, as opposed to brain structures. While the model is largely based on that of Gray, it differs in several important ways, including the major personality dimensions that are postulated, and the greater utilization of human studies in formulating the biological underpinnings of these dimensions.

Zuckerman posits three major personality dimensions, based closely on Eysenck's theory. These are labeled the (1) Psychoticism–Impulsivity–Unsocialized Sensation Seeking (P-Imp-Uss), (2) Neuroticism–Emotional (N-E); and (3) Extraversion–Sociability (E-S) dimensions.

The primary characteristics of a person at one extreme pole of the P-Imp-Uss dimension are high impulsivity and a lack of reflection, a lack of socialization, and disinhibited sensation seeking. The traditional definition of an antisocial personality or psychopath fits this extreme quite closely. Zuckerman (1991) believes the primary problem of the antisocial personality is a failure to inhibit behavior in the presence of cues or anticipation of punishment or negative reinforcement. Stated another way, in approach–avoidance situations, under circumstances in which reward and punishment are both possible, the antisocial personality fails to inhibit their behavior.

Consistent with Strelau's ideas about a strong CNS, Zuckerman posits that the antisocial personality has a greater tolerance for cortical stimulation or excitation. That is, under conditions of high stimulation, the person can continue to function. However, there is a tendency to seek strong levels of stimulation, which when accompanied by impulsivity and a lack of socialization, leads persons extreme on this dimension into antisocial acts.

At the biochemical level, Zuckerman reviews research that demonstrates that persons high on this dimension have deficits in a variety of hormones, neurotransmitters, and enzymes that play important roles in behavioral inhibition. These include cortisol, serotonin, norepinephrine (NE), monoamin-eoxidase (MAO), and dopamine-beta-hydroxylase (DβH).

The case of low levels of platelet MAO seems particularly compelling. Studies of monkeys and humans have shown low levels of MAO to be associated with behavior tapped by the P and E dimensions, although the mechanism by which MAO becomes involved in both dimensions is unclear. However, Zuckerman points out that individual differences in levels of the enzyme have a high heritability, and the enzyme has been associated with behavior very early in life.

A case is also made for higher levels of gonadal hormones, particularly testosterone, in males high on the P-Imp-Uss dimension. High levels of testosterone in prisoners have been related to aggressive criminal records, and in normals to social dominance, sensation seeking, and heterosexual experience. However, Zuckerman points out that the casual connection may be bidirectional; that is, males who are under stress may produce lower levels of testosterone, and males who are successful in competition and in sexual behavior may produce higher levels of testosterone. The connection to successful competition reminds the authors of the old political aphorism that power is the best aphrodisiac.

Persons high on the N-E dimension frequently experience negative emotion, including anxiety, depression, hostility, and anger. Consistent with Tellegen's (1985) theory, Zuckerman believes that the dimension is limited to negative affect, and is not simply a reflection of heightened affect, both negative and positive. Interestingly, he posits that the tendency to frequently experience positive affect is related to the E-S dimension.

Much of the research literature relative to emotionality and N is on anxiety. At the genetic level, Zuckerman's review leads him to conclude that the vulnerability to all anxiety disorders is the subject of polygenetic inheritance, whereas there seems to be a more specific inheritance of panic disorder symptoms, perhaps a dominance pattern.

At the structural level, the amygdala seems to be the most promising structure for control of all kinds of expression of emotionality, including not only fear and anxiety, but aggression as well. Zuckerman seems to believe that the amygdala's function is primarily to assess the emotional significance of stimuli from sensory modalities. There is some evidence that this assessment is carried out before stimuli reach structures involved in cognitive appraisal (cortex or hippocampus).

Neurochemically, Zuckerman's model posits an important place for NE and epinephrine in creating adrenergic arousal, which leads to both aggressive or angry responses, or anxiety, depending on expectations of punishment and other factors. He also hypothesizes that γ-aminobutryric acid (GABA), an inhibitory neurotransmitter that decreases the action of the postsynaptic neuron, plays a central role in the reactions that resulted in emotionality. The idea is that in highly emotional persons, GABA is not present in sufficient quantities, or that its action must be potentiated by some other chemical that is not in sufficient supply. Related to this final point, the hypothesis is entertained that the number of benzodiazepine (BZ) receptors in the nervous system may play a role, because BZ has as its primary function the potentiation of the effects of GABA.

The third macrolevel personality trait discussed by Zuckerman is E-S. Extraverts tends to be more active persons than introverts (in terms of number of activities undertaken and frequency of shifts in type of activity); they also tend to become involved more frequently in social activities. Zuckerman also points out that they are more active in pursuit of reward, and are more optimistic about outcomes.

Zuckerman implicates three structural systems in both activity level and reward sensitivity (which he sees as the central aspects of E-S). All three are mediated by the neurotransmitter DA. These include the caudate nucleus, the substantia nigra, and the ventral segmental area.

He hypothesizes that extraverts may have enough DA-system activity to sustain high levels of activity and reward pursuit. However, a personality group long studied by Zuckerman called sensation seekers are hypothesized to have lower levels of activity in the DA systems. This latter group, which has some characteristics in common with extraverts, are prone to seek highly stimulating and novel environments. Zuckerman posits that they may use such substances and seek intense stimulation to increase activity of these dopaminergic systems. Furthermore, both depressive and stimulant drugs act through the dopaminergic systems, and this may shed light on why sensation seekers abuse both stimulants and depressants.

F. Jerome Kagan

Unlike the other researchers discussed in this chapter, the work of Kagan and colleagues has focused on one cluster of behavioral responses, those related to the initial response to novel or challenging circumstances. Children who demonstrate withdrawal, shyness, timidity, or wariness when they first encounter the unfamiliar are labeled inhibited, while those who are immediately sociable with strangers and who engage in exploratory behavior in unfamiliar contexts are labeled uninhibited. Suomi (1987) and Adamec and Stark-Adamec (1989) have studied similar behavior responses in monkeys and cats, respectively.

Kagan and co-workers (Garcia-Coll, Kagan, & Reznick, 1984; Kagan, Reznick, & Snidman, 1988; Reznick et al., 1986); have conducted a series of studies on two cohorts of children selected to be extreme on initial behavioral inhibition at 21 or 31 months. They have followed some of these children through age 7.5 yr in an effort to determine the stability and the behavioral and physiological correlates of inhibition. Regarding stability, 75% of the children classified as either inhibited or uninhibited at 21 months remained similarly classified at 7.5 yr. Furthermore, in unselected samples (not selected on the basis of extreme inhibition or noninhibition), substantial stability was found for the trait of inhibition. Behavior indices used to determine inhibition included latency to vocalize in an unfamiliar situation, latency to approach an unfamiliar object or person, and time spent in close proximity to mother. Stability of an aggregate of these indices was .40 over a 2-wk period.

Kagan has incorporated a variety of physiological measures into his research on inhibition. Two of the most frequently used measures are heart rate (HR) and heart rate variability (HRV). In general, inhibited children exhibit higher mean HR levels under potentially stressful conditions (not necessarily under baseline conditions, although some studies have found higher HR during sleep for inhibited children) than noninhibited children (e.g., Snidman, 1989).

Furthermore, in several studies conducted by this group, inhibited children were found to have less HRV than the uninhibited (e.g., Snidman, 1989). HRV has been linked to changes in respiratory, vasomotor, or thermoregulatory factors (see Snidman, 1989, for a more extended discussion). One of the most important and well studied of these links is between HRV and inspiration and expiration during normal breathing. During inspiration, the HR increases slightly, and during expiration, it decreases. This relationship is referred to as respiratory sinus arrhythmia.

Typically, HR and HRV are negatively correlated. For example, in the Snidman (1989) study, correlations between HR and HR standard deviation of −.70 and −.56 were obtained from children 31 and 43 months of age, respectively.

A theoretically important issue from a trait conception of inhibition is the stability of these indices. Kagan (1989) reports that these indices were moderately stable over 2 wk in a study of an unselected sample of 32, 14- and 20-month-old subjects. The stability was .64 for HR and .57 for HRV. These two indices were more stable than the behavioral indices of inhibition for the same sample.

Other physiological indices utilized by Kagan and associates included (a) NE activity analysis obtained from urine samples, (b) salivary cortisol assays, (c) pupillary dilation, and (d) variability of vocal utterances under cognitive stress. Reznick et al. (1986) found that at 5.5 yr of age, inhibited children had greater NE activity, and had larger pupillary dilations to cognitive tasks. Kagan et al. (1987) combined all of these indices, including HR and HRV, into a single aggregate of peripheral physiological response, and found that there was a substantial concurrent and postdictive (relationship to a behavior index of inhibition obtained earlier on the sample) relationship with the behavior index of inhibition (.58 concurrent, .70 postdictive).

From these and related physiological findings, Kagan and associates have come to believe that inhibited and uninhibited children differ in the threshold of arousal of limbic structures, with inhibited children exhibiting greater limbic arousal. This variability could be due to quantity of neurotransmitters present, number of pre- or postsynaptic terminals present, or the functioning of uptake mechanisms. Arousal in this context is thought of as the physiological responsivity of these structures in the CNS to physical or psychological stressors.

Some evidence exists in support of this conceptualization. For example, noradrenergic (NA) function has been found to be low in aggressive conduct children (Quay, 1988) and high in children with anxiety disorders (Chaney & Redmond, 1983). High levels of arousal and associated NA function seem to result in hypersensitivity to signals of punishment, over anxious behavior, behavioral inhibition (Rogeness, Javors, Mass, & Macedo, 1990) and interestingly, increased attachment behavior (Steklis & Kling, 1985).

Further, Adamec and Stark-Adamec (1989) have shown that timid, defensive cats show larger evoked potentials in the ventromedial hypothalamus following stimulation of the basomedial amygdala, a finding that is interpreted as indicating greater excitability of this neurological circuit. Kagan (1989), building on this finding and others, hypothesizes that similar effects are characteristic of inhibited children. The line of reasoning is that inhibited children have been shown to exhibit greater sympathetic nervous system activity, and the sympathetic chain arises from the amygdala and hypothalamus.

Kagan (1989) has engaged in some provocative speculation about the etiology of the lower threshold of responsiveness of the hypothalamic–amygdala circuit. He utilizes the concept of preparedness, which is normally applied to interspecies differences, to explain a predisposition for a child to exhibit the physiological and behavioral manifestations of inhibition. In this context, preparedness refers to a state of a biological process, which is present at birth or is produced through developmental maturation, and which "biases the organism to react to events in particular ways" (Kagan, 1989, p. 15). This kind of biological tendency is quite a different explanation than the traditional psychological explanation based on classical or operant conditioning, although the role of conditioning is not denied. Kagan believes that those children who manifest consistently inhibited behavior were in all probability biologically "prepared" for this response, and also experienced a nonbenign environment that strengthened the tendency. Some evidence supports this notion as DβH, an enzyme involved in converting DA to NE, seems to be genetically determined and has been found to be significantly higher in children with anxiety and depression than in children with conduct disorder, where DβH is significantly lower. Thus, there seems to be evidence in support of genetically determined levels of neurotransmitters in the brain that appear in turn to be associated with temperamental predispositions toward anxiety and behavioral inhibition or aggression and associated low levels of anxiety and behavioral disinhibition (Rogeness et al., 1990).

Nonetheless, the heritability studies strongly suggest that genetics cannot account for all of the variance in human behavior. Clearly, some interaction with the environment exists that in turn impacts on biologically driven behavior. Kagan (1989) is sensitive to this point and in addressing it he borrows the concept of kindling from animal researchers who have shown that if a brain structure (such as the amygdala) of an animal is given a short daily burst of electrical stimulation, the structure eventually becomes permanently sensitized. Speculating on the connection of this kind of process to children, it could be hypothesized that the neurochemical circuit from the amygdala to the hypothalamus has been permanently sensitized by repeated encounters with a hostile environment for the fearful and inhibited child, leading to relatively permanent alterations in DβH and the resulting available NE.

IV. SUMMARY AND CONCLUSIONS

There is little doubt that temperament is to a significant degree heritable (Braugart et al., 1992) but there are clear interactions with the environment that appear to induce relatively permanent alterations in both behavior and associated neurotransmitters (Quay, 1988; Rogeness et al., 1990). Theoretically, however, there is considerable divergence of opinion as to the extent and nature of the interactions.

Strelau proposes a regulative theory of temperament (Strelau, 1983). The main idea of the regulative theory of temperament is that temperament, being a product of biological evolution, plays an important role in regulating the interrelations between humans and their environment. Two basic features of the energetic characteristic of behavior are reactivity and activity. Their significance consists mainly in regulating the stimulative value of the surroundings and the individual's own behavior, as a safeguard of his need for stimulation.

Eysenck's theory of personality and temperament is based on the assumption that individuals differ in (1) autonomic nervous system (ANS) reactivity and (2) the degree to which they develop conditioned responses (Eysenck, 1970; Hall & Lindsey, 1978). These individual differences are expressed along personality–temperament dimensions of introversion–extraversion and neuroticism–stability. Conditionability is the underlying trait along the introversion–extraversion dimension. The ANS and central brain structures, such as the limbic system and hypothalamus, are the physiological substrates for emotionality and the ascending reticular-activating system provides the substrates for arousability (Eysenck, 1970). While the theory is rather simplistic (it was generated many decades ago), there is good evidence in support of its basic ideas.

Gray's theory is similar to Eysenck's in many respects. Gray (1976) modified Eysenck's theory by rotating the N and E axes (in two-dimensional space) 45° and renamed the factors impulsivity and anxiety. It was anxiety and impulsivity that Gray (1976) believed to be the causal influences of behavior. Anxiety in Gray's theory was conceptualized as an increasing sensitivity to signals of punishment while impulsivity was seen as the increasing sensitivity to signals of reward. Gray (1976) replaced Eysenck's notion of condition-ability with a fearlessness postulate that states that the greater the degree of trait anxiety, the greater the sensitivity or reactivity to punishment signals and situations.

One can compare Eysenck's theory against those derived from Gray's emotion system model, using anxiety (BIS) and impulsivity (BAS) vs. introversion–extraversion. In this case, the dimensions of anxiety and impulsivity represent the steepest rate of increase in susceptibility to signals of punishment and reward, respectively. Gray's hypothesis predicts that other things being equal, performance and learning will be facilitated or aroused for introverts if aversive rather than appetitive reinforcers are used; while for extraverts performance and learning will be facilitated or aroused if appetitive rather than aversive reinforcers are used. This is contrary to Eysenck's predictions and the research findings.

Zuckerman (1991) hypothesizes that there are three main personality–temperament dimensions: P-Imp-Uss dimension, the E-S dimension, and the N-E dimension. Zuckerman's theory, based heavily on Eysenck's personality dimensions, can be compared to Gray's BAS and BIS systems, respectively. The P-Imp-Uss dimension reflects high-BAS and low-BIS systems, while both the E-S and N-E dimensions most probably have low-BAS and high-BIS systems. Zuckerman's dimensions fit in line perfectly with Eysenck's dimensions.

Finally, Kagan, unlike the other theorists, has only focused on one dimension of temperament: inhibition. His theory and research of inhibited children are very different from what the other theorists propose, yet there do appear to be some similarities between Gray's BIS system and model and Kagan's inhibited children. Not only do the two theories have similar behavioral manifestations, but it also appears that they have similar underlying neurological substrates.

Although from a neuropsychological or neuroanatomical perspective it is very difficult to compare these theories, considering the different types of neuropsychological and behavioral measurements each researcher has used, it is important that further research try to bridge the gaps between the different theories, in order to better understand the neuropsychology of temperament. Although this chapter has attempted to compare and contrast, as well as to clarify the main theories in the neuropsychology of temperament, much research is needed to further compare these different theories. One of the main problems in the field, over the years, has been the lack of cross-communication between the two subareas: neuropsychology, on the one hand, the temperament (behavioral), on the other. Although we have mentioned a few theorists who have crossed the two areas, there still remain many neuropsychologists, as well as temperament theorists, who may share similar interests but do not communicate or know the other exists. We hope that this chapter has not only sparked the interest of neuropsychologists to pursue temperament as a possible manifestation of neurological underpinnings, but also sparked the interest of temperament theorists to further pursue neuropsychological and neuroanatomical explanations of temperament patterns in children.

REFERENCES

Adamec, R. E., & Stark-Adamec, S. (1989). Behavioral inhibition and anxiety: Dispositional, developmental, and neural aspects of the anxious personality of the domestic cat. In J. S. Reznick (Ed.), *Perspectives on behavioral inhibition* (pp. 93–124). Chicago, IL: University of Chicago Press.

Attilli, G. (1990). Successful and disconfirmed children in the peer group: Indices of social competence within an evolutionary perspective. *Human Development, 33,* 238–249.

Bates, J. E. (1989). Concepts and measures of temperament. In G. A. Kohnstamm, J. E. Bates, & M. K. Rothbart (Eds.), *Temperament and childhood* (pp. 3–26). New York: Wiley.

Braugart, J. M., Plomin, R., Defries, J. C., & Fulker, D. W. (1992). Genetic influence on tester-rated infant temperament as assessed by Bayley's Infant Behavior Record: Nonadoptive and adoptive siblings and twins. *Developmental Psychology, 28,* 40–47.

Buss, A., & Plomin, R. (1975). *A temperamental theory of personality development.* New York: Wiley.

Buss, A., & Plomin, R. (1984). *Temperament: Early Personality Traits.* Hillsdale, NJ: Erlbaum.

Cairns, R. B. (1991). Multiple metaphors for a singular idea. *Developmental Psychology, 27,* 23–26.

Canivez, G. L. (1992, March). *Gray's learning theory and ADHD: Psychophysiological implications for research and assessment.* Paper presented at the meeting of the National Association of School Psychologists, Nashville, TN.

Chaney, D. A., & Redmond, D. E., Jr. (1983). Neurobiological mechanisms in human anxiety: Evidence supporting noradrenergic hyperactivity. *Neuropharmacology, 22,* 1531–1536.

Crockenberg, S. B. (1986). Are temperamental differences in babies associated with predictable differences in care-giving? In J. V. Lerner & R. M. Lerner (Eds.), *Temperament and social interaction in infants and children* (pp. 53–74). San Francisco: Jossey-Bass.

Eysenck, H. (1947). *Dimensions of personality.* New York: Praeger.

Eysenck, H. (1959). *Manual of the Maudsley Personality Inventory.* London: University of London Press.

Eysenck, H. J. (1970). *The structure of human personality* (3rd ed.). London: Methuen.

Eysenck, H. J., & Eysenck, S. B. J. (1975). *Manual of the Eysenck Personality Questionnaire (Junior and Adult).* London: Hodder & Stoughton.

Garcia-Coll, C., Kagan, J., & Reznick, J. S. (1984). Behavioral inhibition in young children. *Child Development, 55,* 1005–1019.

Goldsmith, H. H., & Alansky, J. A. (1987). Maternal and infant temperamental predictors of attachment: A meta-analytic review. *Journal of Consulting and Clinical Psychology, 55,* 805–816.

Goldsmith, H. H., Buss, A. H., Plomin, R., Rothbart, M. K., Thomas, A., Chess, S., Hinde, R. A., & McCall, R. B. (1987). What is temperament? Four approaches. *Child Development, 58,* 505–529.

Gottlieb, G. (1991). Experiential canalization of behavioral development: Theory. *Developmental Psychology, 27,* 4–13.

Gray, J. A. (1976). The behavioral inhibition system: A possible substrate for anxiety. In M. P.

Feldman & A. M. Bradhurst (Eds.), *Theoretical and experimental bases of behavior modification, Vol. 3* (pp. 3–41). New York: Wiley.

Gray, J. A. (1991). The neuropsychology of temperament. In J. Strelau & A. Angleitner (Eds.), *Explorations in temperament: International perspective on theory and measurement*. New York: Plenum Press.

Groves, P. M. (1983). A theory of the functional organization of the neostriatum and the neostriatal control of voluntary movement. *Brain Research Reviews, 5,* 109–132.

Hall, G. S., & Lindsey, G. (1978). *Theories of personality*. New York: Wiley.

Huttunen, M. O., & Nyman, G. (1982). On the continuity, change and clinical value of infant temperament in a prospective epidemiological study. In M. Rutter (Ed.), *Temperament differences in infants and young children* (pp. 240–247). (Ciba Foundation symposium 89). Pitman: London.

Kagan, J. (1989). The concept of behavioral inhibition to the unfamiliar. In J. S. Reznick (Ed.), *Perspectives on behavioral inhibition* (pp. 1–24). Chicago, IL: University of Chicago Press.

Kagan, J., Reznick, S., & Snidman, N. (1988). Biological bases of childhood shyness. *Science, 240,* 167–171.

Keogh, B. K. (1982). Temperament: An individual difference of importance in intervention programs. *Topics in Early Childhood Special Education, 2,* 25–31.

Lou, H. C., Henriksen, L., & Bruhn, P. (1984). Focal cerebral hypoperfusion in children with dysphasia and/or attention deficit disorder. *Archives of Neurology, 41,* 825–829.

Lou, H. C., Henriksen, L., Bruhn, P., Børner, H., & Nielsen, J. B. (1989). Striatal dysfunction in attention deficit and hyperkinetic disorder. *Archives of Neurology, 46,* 48–52.

Martin, R. P. (1989). Temperament and educational outcomes: Implications for underachievement and learning disabilities. In W. Carey, and S. McDevitt (Eds.), *Temperament as a risk factor for children* (pp. 104–115). Amsterdam: Swets.

Martin, R. P., Wisenbaker, J., & Huttunen, M. O. (in press). The factor structure of instruments based on the Chess-Thomas model of temperament: Implications for the Big Five. In C. F. Halverson, G. Kohnstamm, & R. Martin (Eds.), *The developing structure of temperament personality from infancy to adulthood*. Hillsdale, NJ: Erlbaum.

Martin, R. P., Wisenbaker, J., Mathews-Morgon, J., Holbrook, J., Hooper, S., & Spalding, J. (1986). Stability of teacher ratings of temperament at 6 and 12 months. *Journal of Abnormal Child Psychology, 14,* 167–179.

Matheny, A. P., Jr. (1980). Bayley's Infant Behavior Record: Behavioral components and twin analysis. *Child Development, 51,* 1157–1167.

Matier, K., Halperin, J. M., Sharma, V., Newcorn, J. H., & Sathaye, N. (1992). Methylphenidate response in aggressive and nonaggressive ADHD children: Distinctions on laboratory measures of symptoms. *Journal of the American Academy of Child and Adolescent Psychiatry, 31,* 219–225.

Milich, R. (1993, February). *Disinhibition and underlying processes in hyperactive and aggressive adolescents*. Paper presented at the meeting of the Society for Research in Child & Adolescent Psychopathology. Santa Fe, NM.

Plomin, R. (1990a). The role of inheritance in behavior. *Science, 248,* 183–188.

Plomin, R. (1990b). *Nature and nurture*. Pacific Grove, CA: Brooks/Cole.

Plomin, R., & Rende, R. (1991). Human behavioral genetics. *Annual Review of Psychology, 42,* 161–190.

Presley, R., & Martin, R. P. (in press). Toward a structure of childhood temperament: A factor analysis of the TABC. *Journal of Personality*.

Quay, H. C. (1988). The behavioral reward and inhibition system in childhood behavior disorder. In L. M. Bloomingdale (Ed.), *Attention deficit disorder, Vol. II* (pp. 177–186). New York: Pergamon Press.

Reznick, J. S., Kagan, J., Snidman, N., Gersten, M., Baak, K., & Rosenberg, A. (1986). Inhibited and uninhibited behavior: A follow-up study. *Child Development, 57,* 660–680.

Rogeness, G. A., Javors, M. A., Mass, J. W., & Macedo, C. A. (1990). Catecholamines and

diagnosis in children. *Journal of the American Academy of Child and Adolescent Psychiatry,*
29, 234–241.

Rothbart, M. K. (1989). Temperament and development. In G. A. Kohnstamm, J. E. Bates, &
M. K. Rothbart (Eds.), *Temperament in childhood* (pp. 187–248). Chichester, England:
Wiley.

Satterfield, J. H. (1978). The hyperactive child syndrome: A precursor of adult psychopathy? In
R. D. Hare & D. Schalling (Eds.), *Psychopathic behavior: Approaches to research* (pp. 329–
346). New York: Wiley.

Scarr, S., & McCartney, K. (1983). How people make their own environments: A theory of
genotype environment correlations. *Child Development, 54,* 424–435.

Snidman, N. (1989). Behavioral inhibition and sympathetic influence on the cardiovascular
system. In J. S. Reznick (Ed.), *Perspectives on behavioral inhibition* (pp. 51–70). Chicago, IL:
University of Chicago Press.

Steklis, H. D., & Kling, A. (1985). Neurobiology of affiliative behavior in nonhuman primates.
In M. Reite & T. Field (Eds.), *The Psychobiology of Attachment.* New York: Academic Press.

Strelau, J. (1983). *Temperament personality activity.* New York: Academic Press.

Strelau, J. (1985). Temperament and personality: Pavlov and beyond. In J. Strelau, F. H. Farley,
& A. Gale (Eds.), *The biological bases of personality and behavior: Theories, measurement
techniques, and development,* Vol. 1 (pp. 25–39). New York: Hemisphere Publishing Corporation.

Suomi, S. J. (1987). Genetic and maternal contributions to individual development. In N. A.
Krasnegor, E. M. Blass, M. A. Hofer, & W. P. Smotherman (Eds.), *Perinatal development: A
psychobiological perspective* (pp. 397–420). San Diego: Academic Press.

Swanson, J., Kinsbourne, M., Roberts, W., & Zucker, K. (1978). Time-response analysis of the
effect of stimulant medication on the learning ability of children referred for hyperactivity.
Pediatrics, 61, 21–24.

Swerdlow, N. R., & Koob, G. F. (1987). Dopamine, schizophrenia, mania, and depression:
Toward a unified hypothesis of cortico-striato-pallido-thalamic function. *Behavioral and
Brain Sciences, 10,* 215–217.

Tellegen, A. (1985). Structures of mood and personality and their relevance to assessing anxiety,
with emphasis on self-report. In A. H. Tuma & J. D. Maser (Eds.), *Anxiety and the anxiety
disorders* (pp. 681–716). Hillsdale, NJ: Erlbaum.

Teplov, B. M., & Nebylitsyn, V. D. (1963). The study of basic properties of the nervous system
and their significance in psychology of individual differences. *Voprosy Psikhologii, 5,* 38–47.

Thomas, A., & Chess, S. (1977). *Temperament and development.* New York: Bruner/Mazel.

Thomas, A., Chess, S., & Birch, H. (1968). *Temperament and behavioral disorders of childhood.*
New York: New York University Press.

Zuckerman, M. (1991). *Psychobiology of personality.* Cambridge, England: Cambridge University Press.

Author Index

Subject Index